FM 25-6

US WAR DEPARTMENT FIELD MANUAL

DOG TRANSPORTATION

1944

WORLD WAR II

CIVILIAN REFERENCE EDITION

UNABRIDGED HISTORIC MANUAL ON SNOW DOG SLEDDING TEAMS, TRAINING, KENNEL MANAGEMENT, WINTER TRAVEL EQUIPMENT, AND CARE FOR ANIMALS

U.S. WAR DEPARTMENT

Doublebit Press

www.doublebitpress.com | Cherry, IL, USA

Original content under the public domain; unrestricted for civilian distribution.
Originally published in 1944 by the U.S. War Department.

A Part of the Military Outdoors Skills Series

Doublebit Press Civilian Reference Edition ISBNs
Hardcover: 978-1643892009
Paperback: 978-1643892016

The Military Outdoors Skills Series

Historic Field Manuals and Military Guides on Outdoors Skills and Travel

Military manuals contain essential knowledge about outdoors life, thriving while in the field, and self-sufficiency. Unfortunately, many great military books, field manuals, and technical guides over the years have become less available and harder to find. These have either been rescinded by the armed forces or are otherwise out of print due to their age. This does not mean that these manuals are worthless or "out of date" – in fact, the opposite is true! It is true that the US Military frequently updates its manuals as its protocols frequently change based on the current times and combat situations that our armed services face. However, the knowledge about the outdoors over the entire history of military publications is timeless!

By publishing the **Military Outdoors Skills Series**, it is our goal at Doublebit Press to do what we can to preserve and share valuable military works that hold timeless history about outdoors life, navigation, and survival. These books include official unrestricted texts such as army field manuals (the FM series), technical manuals (the TM series), and other military books from the Air Force, Navy, and texts from before 1900. Through remastered reprint editions of military handbooks and field manuals, outdoors enthusiasts, bushcrafters, hunters, scouts, campers, survivalists, nature lore experts, and military historians can preserve the time-tested knowledge and history about the skills and institutional procedures that were learned through hard lessons and training by the U.S. Military and our expert soldiers.

This book is an important contribution to self-sufficiency literature and has important historical and collector value toward preserving these traditions.

Although the editors at Doublebit Press are thrilled to have comfortable experiences in the woods and love our high-tech and lightweight equipment, we are also realizing that the basic skills taught by the old experts are more essential than ever as our culture becomes more and more hooked on digital technology. We don't want to risk forgetting the important steps, skills, or building blocks involved with thriving in the outdoors. This Civilian Reference Edition historic reprint is a part of a collection of military handbooks and field manuals that are essential contributions to the American self-sufficiency tradition despite originating with the military. In the most basic sense, these books are the collection of experiences by the great experts of outdoors skills and self-sufficiency life: our countless expert soldiers who learned to thrive in the backwoods, deserts, extreme cold environments, and jungles of the world.

With technology playing a major role in everyday life, sometimes we need to take a step back in time to find those basic building blocks used for gaining mastery – the things that we have luckily not completely lost and has been recorded in books over the last two centuries. These history hasn't been forgotten, it's just been shelved. *It's time to unshelve this history once again and reclaim the lost knowledge of self-sufficiency.*

Based on this commitment to preserving our self-sufficiency heritage, we have taken great pride in publishing this book as a complete, unabridged and original work. We hope it is worthy of both study and collection by outdoors folk in the modern era of self-sufficiency and traditional skills life.

Unlike many other photocopy reproductions of classic books that are common on the market, this historic Civilian Reference Edition does not simply place poor photography of old texts on our pages and use error-prone optical scanning or computer-generated text. We want our work to speak for itself, and reflect the quality demanded by our customers who spend their hard-earned money. With this in mind, each Historic Edition book that has been chosen for publication is carefully remastered from original print books, *with the Doublebit Civilian Reference Edition printed and laid out in the exact way that it was presented at its original publication*. We provide a beautiful, memorable experience that is as true to the original text as best as possible, but with the aid of modern technology to make as beautiful a reading experience as possible for books that are typically over a century old. Military historians and outdoors enthusiasts alike are sure to appreciate the care to preserve this work!

Because of its age and because it is presented in its original form, the book may contain misspellings, inking errors, and other print blemishes that were common for the age. However, these are exactly the things that we feel give the book its character, which we preserved in this Historic Edition. During digitization, we ensured that each illustration in the text was clean and sharp with the least amount of loss from being copied and digitized as possible. Full-page plate illustrations are presented as they were found, often including the extra blank page that was often behind a plate. For the covers, we use the original cover design to give the book its original feel. We are sure you'll appreciate the fine touches and attention to detail that your historic Civilian Reference Edition has to offer.

For outdoors, self-sufficiency, and military history enthusiasts who demand the best from their equipment, the Doublebit Press Civilian Reference Edition reprint of this military manual was made with you in mind. Both important and minor details have equally both been accounted for by our publishing staff, down to the cover, font, layout, and images. It is the goal of Doublebit Civilian Reference Edition series to preserve outdoors and self-sufficiency heritage, but also be cherished as collectible pieces, worthy of collection in any library and that can be passed to future generations.

WAR DEPARTMENT FIELD MANUAL

FM 25-6

This manual supersedes FM 25–6, Dog Team Transportation, 4 January 1941.

DOG TRANSPORTATION

WAR DEPARTMENT 19 AUGUST 1944

United States Government Printing Office

Washington: 1944

WAR DEPARTMENT,

WASHINGTON 25, D. C., 19 August 1944.

FM 25–6, Dog Transportation, is published for the information and guidance of all concerned.

[A.G. 300.7 (17 Apr 44).]

BY ORDER OF THE SECRETARY OF WAR:

G. C. MARSHALL,

Chief of Staff.

OFFICIAL:

J. A. ULIO,

Major General,

The Adjutant General.

DISTRIBUTION:

War Dog Reception and Tng Centers (200).

(For explanation of symbols see FM 21–6.)

CONTENTS

This manual supersedes FM 25–6, Dog Team Transportation, 4 January 1941.

CHAPTER 1

SLEDGE DOGS

Section I. USE OF SLEDGE DOGS

1. GENERAL. a. In a world where the forms of transportation have been brought to a high state of perfection, where the airplane is fast supplanting other methods of travel, there is still a place for the use of sledge dogs. Dog transportation will perhaps always remain supreme for certain kinds of work in Arctic and sub-Arctic areas. Where matériel is to be transported over known, uniform, flat and hard-packed snow or ice surfaces, specially designed motor vehicles have proved more efficient than dogs. But in regions where there is no such uniformity—where the route to be traversed includes pack-ice on the sea or crevassed and mountainous areas—only the sledge dog can be used with success and safety. Both motorized units and sledge dogs have their limitations. The former can carry heavy loads with considerable speed but are limited as to the type of terrain on which they can operate effectively. Sledge dogs provide a surer method of travel over varied types of terrain but are slower and more limited as to the load they can carry.

b. Sledge dogs are the only reliable means of transportation over land and sea-ice in the Arctic regions. Eskimos, trappers, traders, explorers, police, and missionaries, all depend on and use dogs for travel. The Army has already employed dogs successfully in airplane salvage operations, on rescue missions, and for normal travel. Ground troops find dogs useful and often indispensable in transporting supplies and equipment, as well as for ambulance service over terrain on which it is impossible for motor vehicles to operate. In World War I, the French used approximately 400 Canadian sledge dogs in the Vosges Mountains. They operated in snow so deep that it nearly brought a halt to all activities in the area. It is recorded that one section of these dogs took but 4 days to haul 90 tons of ammunition up to a battery, which men, horses, and mules had vainly struggled to reach for a fortnight. When the snow melted, the dogs were harnessed to a small-gauge light railway and proved more useful than horses in that precipitous country.

2. NORMAL OPERATIONS. Dog teams may be classified into messenger and freight teams, although at an Arctic outpost, with its limited number of available dogs, any such clear-cut separation would hardly be practicable.

a. Messenger teams. Fast teams are used when speedy communication is necessary, and for emergencies such as relief expeditions and rapid transportation of the sick or of medical supplies. Rescue parties for airplanes which have made forced landings in isolated places usually employ fast dog teams. A good general rule to follow in the loading of messenger teams is that loads should not average more than 50 pounds per dog.

b. Freight teams. Freight teams, as the name implies, are used for hauling equipment and supplies. The slower, heavier dogs, each capable of pulling freight equal to or greater than his own weight (100 pounds or more), should be used for this purpose. Under normal sledging conditions and on a relatively flat surface, a freight team pulling a heavy load (½ ton or over) can travel 3 to 5 miles an hour. The daily average should be about 3 miles an hour, including the necessary halts for rest.

3. RESCUE OPERATIONS. Rescues of Army airplane crews forced down in the Arctic have been and will increasingly be accomplished through the use of dogs, or through the combined use of planes, motor sledges, and dogs. Dogs have generally proved most reliable because of their ability to surmount the obstacles of uncertain weather, treacherous crevasses, and rough and hilly terrain. In Alaska and Greenland, many rescues which would have been impossible by any other means have been made by dog sledge. Such rescue operations require the best-trained, as well as the fastest and best-pulling dogs.

Section II. DOG TYPES AND CHARACTERISTICS

4. GENERAL. a. A sledge dog is commonly referred to as an Eskimo dog or "husky." The term "Eskimo" comes from the fact that native Eskimo tribes use the dog more widely than any other people. There is a general belief that the sledge dog is more closely related to the dog tribe's common ancestor, the wolf, than the common domesticated dog. However, there is much difference of opinion on this subject. Both the domesticated dog and the wolf are classified in the same genus, *Canis,* of the animal kingdom, and there are a great many records of cross-breeding between them. There can be little doubt that the Eskimo practice of allowing dogs to run loose in the summer has resulted in interbreeding with the Arctic wolf. Where wolves and "semidomesticated" sledge dogs range together, it is reasonable to expect some interbreeding. However, the wolf strain is negligible in sledge dogs bred apart from wolves.

b. The reputation of sledge dogs for savagery is ill-founded. Treated with understanding, in a kindly manner, they are as friendly to man as any other type of dog. Among themselves they exhibit a considerable tendency to fight; and although they are excellent fighters, they often exhibit cowardly traits. They may attack a dog who is off his guard, or a group may attack a lone dog. These tendencies make it imperative to secure the dogs at all times. Failure to do so leads invariably to dog fights and consequent casualties. Such a situation might have serious consequences at a time when dogs are badly needed.

5. BREEDS. a. The American Kennel Club recognizes four distinct breeds of sledge dogs. The Alaskan Malemute, the Eskimo, the Siberian Husky, and the Samoyede. All are popularly called "husky", but this name applies properly only to the Siberian. All these breeds are used in the Arctic; all are characterized by a wolflike appearance, thick fur, bushy plumelike tail (carried differently

Figure 1. Characteristic type of sledge dog.

from the wolf's, however), tough hairy feet, and a long-drawn mournful howl, rather than a bark. Common colors are wolf-gray, black, white, or a combination of all three.

b. The majority of sledge dogs are a cross between

true types and many other breeds, such as the St. Bernard, the German shepherd, the pointer, the labrador, the collie, and even the greyhound. However, within the past 15 years, Malemute, Eskimo, and Siberian Husky dog clubs have been organized in the United States, each with the objective of improving the breed. Any long-range, Army sledge dog selective breeding program should attempt to improve the strain, whatever breed may be selected. The following descriptions are recognized as the standard type for each breed:

(1) *The Malemute.* This is the native sledge dog of Alaska, developed by the Alaskan Malemute people (a group of Eskimo tribes) for draft purposes. It is one of the heaviest and most powerful of all sledge dogs, and has an excellent record in the history of Arctic and Antarctic exploration.

(*a*) *General appearance and characteristics.* The Malemute is a large dog with a strong, compact body; it stands well over 30 inches from the pads of the feet to the top of the head and has an appearance of strength and stamina. Outstanding characteristics are a deep chest, a proud carriage, an erect head, and an alert look. A distinguishing feature consists of a caplike marking over the head, with the rest of the face a different solid color (usually grayish white), or a "mask" that sets off the eyes.

(*b*) *Head.* The skull is broad between the ears, gradually narrowing to the eyes. The muzzle is somewhat stubby. The nose is black and the lips are close-fitting, the upper and lower jaws being broad and set with large teeth. The eyes are almond-shaped, moderately large, and set obliquely in the skull, giving the Malemute a somewhat wolklife appearance. The ears are set wide apart on the outside back edges of the top of the skull; the upper half of the ear is triangular but slightly rounded at the tip; the ears are generally held erect, giving the Malemute its alert look.

(*c*) *Shoulders, legs, and feet.* The shoulders are moderately sloping. The forelegs are heavily boned and muscled. The large compact feet have thick tough pads and well-arched toes, with a protective growth of hair between them. The pads are broad and not long. The hind legs are broad and powerfully muscled through the thighs, and have moderately bent stifles and strong, broad hock joints. Viewed from the rear, the hind legs do not appear bowed. The legs are of medium length.

(*d*) *Coat.* The coat is dense and coarse but not long; the under coat is thick, oily, and woolly; the outer coat is rather coarse and stands out. The fur is thicker around the neck and inside the thighs. The fur of the Malemute is coarser than that of the Siberian Husky. The usual colors are wolf-gray or black and white; color variations and markings are occasionally found. Markings on the face are either caplike or masklike.

(*e*) *Tail.* The tail is well-furred and carried over the back in a plumelike manner.

(*f*) *Height and weight.* The Malemute male averages 22 to 25 inches from the top of the back to the pads; the bitch averages 20 to 23 inches. The adult male averages 65 to 85 pounds in weight; the bitch weighs approximately 10 percent less.

(2) *The Eskimo.* This sledge dog is closely akin to the Malemute and Siberian Husky. Although found throughout the Arctic, it is generally known as the type of dog used in Greenland and the eastern Canadian arctic regions. Most of the dogs used by Admiral Peary when he sledged to the North Pole were of this type. Amundsen used them in the Antarctic, and all of the Byrd and United States Antarctic expeditions have used Eskimo dogs and found them very satisfactory.

(*a*) *General appearance.* The Eskimo has a powerful body, a keen expression, erect ears, slanting eyes, a tightly curled tail, thick fur, a broad chest, and legs of medium length.

(*b*) *Head.* The head is well-proportioned, broad, and wedge-shaped. The strong flat skull has powerful jaws. The nose is large and black or brown. The lips are black or brown. The eyes are small and deep. The ears are short and firm and are carried erect and forward.

(*c*) *Shoulders, legs, and feet.* The neck is short,

heavy, and very muscular. The forequarters are broad, big-boned and muscular; the forelegs are straight. The chest is deep and broad, with the ribs well-sprung. The hindquarters are of medium length, heavy, and very muscular, with the stifles well bent. The large paws have thick pads which are well protected with hair.

(*d*) *Coat.* This breed probably has the heaviest coat of any sledge dog. The hair varies from three to six inches in length and is very thick. Nearly all gradations of black, white, black and white, wolf-gray, blue-gray, and all shades of tan and buff are found. The most common markings are the following: white bodies with coal-black heads; white heads with coal-black bodies; or all white, with double spinal markings of silver gray.

(*e*) *Tail.* The tail is large and bushy and is carried up or curled over the back.

(*f*) *Height and weight.* Similar to the Malemute.

(3) *The Siberian Husky.* This breed came originally from northeastern Siberia and has probably been bred truer to type, over a period of time, than either the Malemute or the Eskimo. It is the fastest of the four breeds and holds most of the racing records. Although lighter in weight than the Malemute or Eskimo, it is one of the hardiest types, and its record in both the Arctic and Antarctic is excellent. When the emphasis is on speed and endurance, the Siberian is often favored; for heavy pulling and endurance, the Malemute and Eskimo are generally preferred.

(*a*) *General appearance.* The Siberian is an alert, gracefully built, medium-size dog, with a strong, compact body. It has a deep, strong chest and well-muscled hindquarters. The head has a foxlike appearance. The tail is carried over the back.

(*b*) *Head.* The head is of medium size in proportion to the body; the width between the ears is medium or narrow, gradually tapering to the eyes and moderately rounded. The muzzle is of medium length and slightly more pointed than the Malemute's or Eskimo's. The nose is light brown (black, in white Siberians). The eyes are set somewhat obliquely in the skull and are generally either brown or blue. The ears are small or medium in size, set high on the head, and moderately pointed.

(*c*) *Body.* The chest is deep, strong, and not very broad; the ribs are well-arched and deep; the shoulders are powerful. The back is of medium length. The loins are well-muscled and slightly arched.

(*d*) *Legs.* The legs are well-muscled and straight. The feet are oblong in shape and not as broad as the Eskimo's or Malemute's. The pads are tough, thickly cushioned, and well-furred between the toes.

(*e*) *Coat.* The coat of the Siberian Husky may be a mixture of black, gray, and white, but never with any trace of red, brown, or yellow. The undercoat is thick, soft, and downy next to the skin. The outer coat of guard hairs lies close to the body and, compared to the outer coat of the Eskimo and Malemute, is thick and soft-textured.

(*f*) *Height and weight.* Both height and weight distinguish this breed from the others. The male averages 21 to 23½ inches in shoulder height and 50 to 60 pounds in weight. Bitches stand 20 to 22 inches high and weigh 40 to 50 pounds.

(4) *The Samoyede.* This breed was developed by the Samoyede tribe in Russian Siberia. It is most nearly akin to the primitive dog and is supposed to have no admixture of wolf. The species is very gentle and friendly, and has been developed for companionship and for herding reindeer as well as for sledging. Pure strains are all white. The average weight of the Samoyede is slightly less than that of the other breeds. The Samoyede is not highly regarded for hard sledging and is not as efficient a sledge dog as those of the other three breeds.

6. AGE. Sledge dogs generally have a shorter span of life than the ordinary domesticated dog. Nine years is about the average life span. Hard work and extremely cold weather in part account for this relatively short life. The training of sledge dogs should begin at about 6 months. With proper training, the dog can take its place in the team at 1 year. Between the ages of 3 and 6 years, the dog should be at its peak for hard work.

7. DOG SPEEDS. The record for speed and endurance among sledge dogs was made in the famous serum dash from Nenana to Nome, Alaska, in the winter of 1925. By relays of teams and drivers, the 665-mile distance was covered in 5½ days, an average of 125 miles a day. This was accomplished under the most adverse weather conditions. The dogs used were mostly Malemutes and Siberian Huskies. The record for the All-Alaska Sweepstakes, a nonstop race of 408 miles, was made by a team of Siberian Huskies which covered the distance in 74 hours and 17 minutes. (See par. 67 for average rates and distances of travel.)

CHAPTER 2

SLEDGING EQUIPMENT

Section I. SLEDGES

8. GENERAL. A wide variety of sledges has been developed over the years, and certain distinctive types have emerged. In addition, many types of sledges may be built or improvised. The paragraphs below describe those basic sledges which have proved their worth for various purposes and which therefore are worthy of consideration for use by Army sledgers. All sledges should be constructed of the best materials available. Runners should be of hickory with upper parts of ash; oak may also be used. All wood should be close-grained and properly seasoned to minimize any tendency to split. It is advisable to paint all wooden parts except the runners with white paint which prevents snow from sticking and makes excellent camouflage. To fit sledges for all-around sledging operations, they should be equipped with steel shoes and a bridle rope, as well as with a canvas tank and lashing ropes.

a. Nansen double-end freight sledge. The Nansen sledge, with some modifications, is one of the sledges used with Army dog teams (fig. 2). The principal changes in the original model include the incorporation of a gee-pole and boards on the undersides of the bridges. The boards help the sledge to act as a toboggan in deep snow, but their main function is to protect the bridges when passing over sharp, jagged ice.

b. Antarctic sledge. This is an excellent freight type (modified Nansen) sledge, which has been used by Americans in the Antarctic. (See figs. 3, 6, 12, 18, 26, 32,

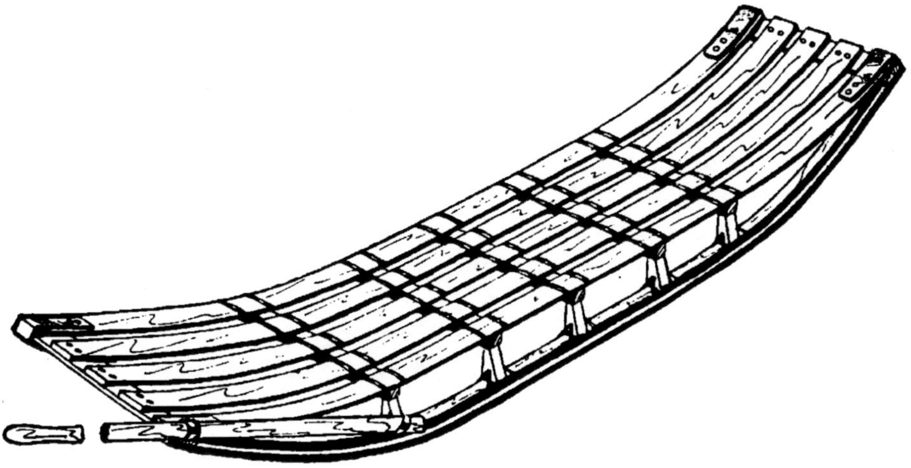

Figure 2. Nansen double-end freight sledge.

and 34.) It is the lightest type of sledge; with the arch construction of the Czegka bridge it will carry the largest load in proportion to its weight. It was primarily designed for sledging on Antarctic snow and ice, which is fairly smooth, but can easily be adapted for use in Arctic regions by equipping it with steel shoes, and by lashing ½-inch boards along the under sides of the bridges for protection against jagged ice. This sledge is an excellent

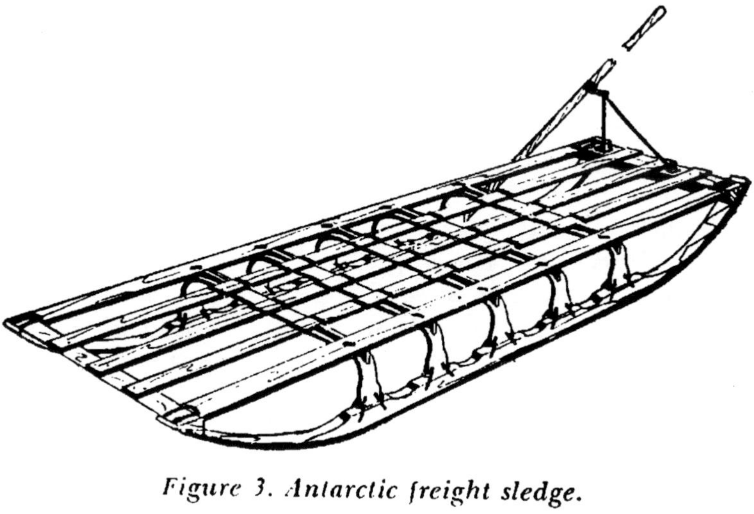

Figure 3. Antarctic freight sledge.

type for use in areas such as the Greenland Ice Cap. It should be equipped with a gee-pole (par. 9) and a sledge tank (par. 11) to fit it for trail operation.

c. Trailer sledge. Any type of sledge may be used as a trailer sledge, but the Antarctic or the Nansen type is best in most cases. It is not necessary to equip a trailer sledge with a gee-pole; a sledge tank, however, should be provided. The only special equipment necessary is a trailer hitch or bridle (fig. 18).

d. Basket type sledge. This sledge is the most widely used of any of the various types, and will best serve as a

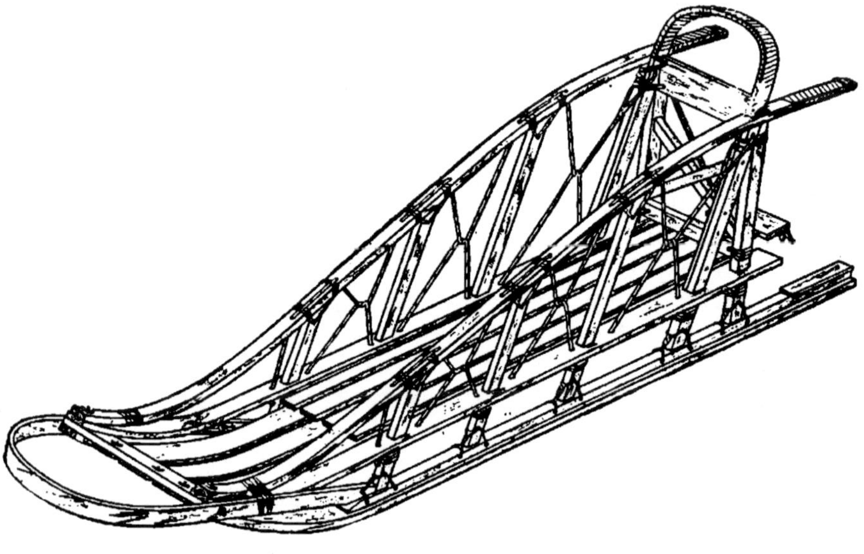

Figure 4. Basket type sledge.

passenger sledge, a training sledge, and an all-purpose, light-freight sledge. (See figs. 4, 25, 31, and 33.) There is a 10½-foot freight sledge and a lighter 9½-foot messenger sledge in this style. The 10½-foot sledge is designed for a normal load of 400 pounds and a capacity load of 750 pounds; it weighs approximately 100 pounds. In addition, there is a convertible, collapsible basket type sledge of plywood construction. This can be used as a toboggan, a freight sledge, or a basket type sledge. It is designed for a normal 750-pound load and a capacity load of 1,200 pounds. While other types of freight sledges can carry heavier loads, the basket sledge is easier to control and brake. It can be used with a gee-pole, although common practice is to travel in the rear of this type of sledge. The lashed sides make it especially adaptable for ambulance use in rescue work. It is a heavier sledge than the Antarctic type. In this as well as other types of sledges, the joints should be lashed in order to provide resiliency when passing over uneven surfaces.

e. "Komatik" or Greenland sledge. Eastern Canadian Arctic and Greenland Eskimos use a "komatik"

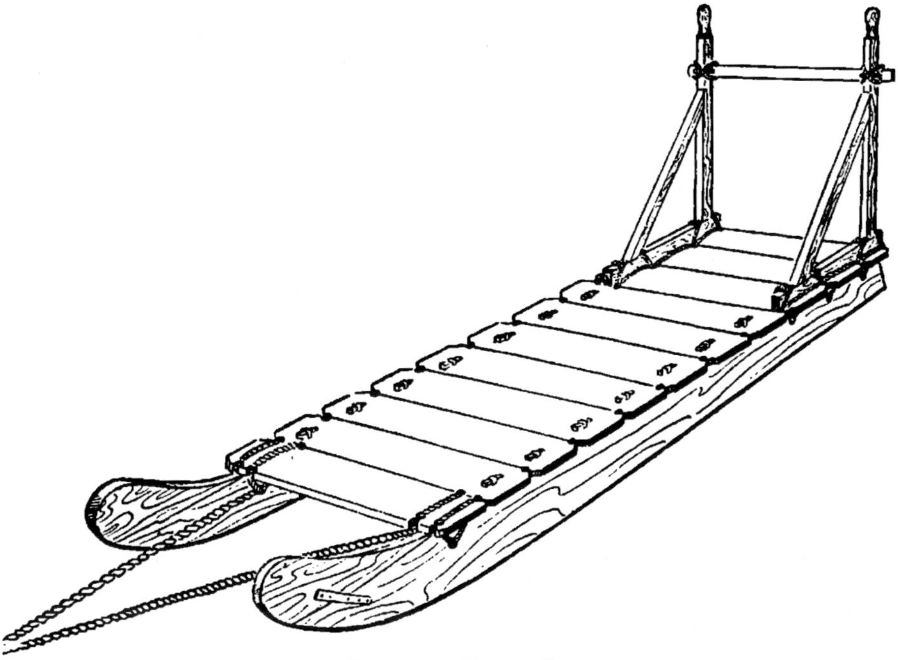

Figure 5. Komatik.

or sledge, varying in length from 6 to 30 feet, and in width from 15 to 30 inches; 14 feet is the average length, and considerable width seems to be favored.

(1) Variations in the dimensions of the komatik seem to be determined both by materials available and by the ideas of the builders. The experiences of white men who have sledged in Greenland demonstrate that a komatik

built along the lines described below is very satisfactory for general use. (Fig. 5; see also fig. 29.) A good komatik should be constructed entirely of a well-seasoned ash, and painted or varnished to guard against waterlogging. The runners are about 9 or 10 inches deep and 3/4 inch thick; they slant out from the sledge platform at an angle of 110°. They are covered with well-polished spring steel, 1 1/4 inches wide and 3/16 inch thick. The steel shoeing tapers at the forward extremity to curve around the upturned, blunt-nosed projections, which are fastened (either with screws or glue) to the runners in front. The platform of the sledge is made of 10 or 11 cross-boards. To allow some "give" in travel over rough terrain, the front board and the three boards at the rear of the sledge are secured to the runners with rope, rawhide (babbiche or babish) thongs of bearhide or sealskin, which is looped through holes in the runners. (About 7 feet of cord or rawhide is required for each lashing.) Each of the six boards between is let into the runners with a strip of steel, which is screwed in place and projects above the board in a square loop; a wedge inserted in this loop "locks" the board. To prevent further splaying of the runners, a supporting wood bar is fitted just under the slightly projecting ends of the crosspieces along the outside of each runner. The usual komatik is approximately 2 1/2 feet wide between the wedge-loops and about 5 1/3 feet long. However, these dimensions are by no means fixed. To form a pulling rope, an 8-foot rope is inserted in the front holes of the runners and pulled forward from the inside. The rope is then knotted beyond the bumper-like ends. The trace-line is attached to this V-shaped "bridle." The handle bars (at the rear of the sledge) are buttressed and lashed to the runners with manila cord threaded through the first and third holes in the runners at the rear.

(2) Because of their depth, weight, and runner width, native komatiks are not suitable for use in extremely soft snow, nor are they advantageous for sledging in the interior where surfaces are comparatively smooth and level. However, when it is necessary to sledge with heavy

loads over rough and jagged sea-ice, the komatik is one of the few types capable of standing up under the strain. Construction of a komatik is comparatively simple, and it is possible to build one at any northern Army base, if the terrain requires its use.

f. Toboggans. Toboggans are very effective as ambulances, for freighting where snow is soft and deep, and particularly in wooded areas. The Canadian Indians use toboggans extensively. Toboggans are much lighter than most sledges and will stand a great deal of rough treatment.

(1) The main objection to the use of toboggans is that they cannot carry such heavy loads as sledges, because of their much larger friction surface. An additional disadvantage is that the load will become wet if it is necessary to sledge through slush or water.

(2) Toboggans should be at least 9 feet long. In wooded country a 17-inch width will best conform to snowshoe tracks, and to the tracks of dogs in single tandem. A 22-inch width is best for ambulance work in open country. In the wide type, the outside and center strips should be of thicker material so as to form a low runner and prevent side-slipping.

g. Siwash toboggan. This type, with the single tandem hitch, is commonly used by the Indians on the narrow trails in the wooded areas of Yukon Territory. It is usually 17 inches wide and weighs about 50 pounds. This toboggan is generally made of two or three boards held together by four wood cleats. The railing handle and the uprights are fastened to the two rear cleats. An iron hinge brake is attached to the rear. An eyebolt is placed in the front edge of each bow or upturn; ropes are fastened to these and stretched to the rear upright. Canvas is then sewed around these ropes, drawn down each side of the toboggan, and fastened to the toboggan floor. The canvas "tank" thus formed prevents the load from falling off the toboggan. A light toboggan of this style is excellent for ambulance work in wooded regions. The Mackenzie River type toboggan is readily adaptable to this style.

h. Training sledges. A training sledge must allow the driver to exert complete control over his team. This can be accomplished through the use of either weight or brakes. Brakes are by far the most satisfactory means of control because they permit adjustments to be made for changes in dog power. There is no standard issue training sledge, but several different types may be improvised. Any type used must be constructed to withstand hard usage.

(1) *Stoneboat.* This is a sledgelike structure similar to that employed by farmers for hauling rocks from a field. The runners consist of two tree trunks or poles, 6 to 8 inches in diameter, 6 feet long, and turned up at one end. Across these, 2-inch planking may be spiked. The "sledge" should be 3 to 5 feet wide. The bridle can be fastened to the sledge by drilling a hole through the upturned part of each runner, threading the rope through each hole, and knotting the ends. This type of training sledge may be used on bare ground, where its weight will provide for control of the dogs.

(2) *"Jeep" sledge.* Training will often have to be carried on when the ground is barren of snow, and in this case a wheeled vehicle can best be used. A discarded jeep, Austin, or Model T Ford can easily be transformed into an excellent training sledge. The vehicle should be stripped of all unnecessary equipment, including the motor; it is essential to retain only the wheels, chassis, seats, steering apparatus, and brakes. It can then be used on bare ground, hitched to from five to fifteen dogs. The use of mechanical brakes gives the driver better control over his dogs than can be obtained with any type of sledge.

9. SLEDGE ACCESSORIES. a. Gee-pole. This device was originally used in the Yukon region and has been standard equipment on the sledges of all the Byrd and United States Antarctic expeditions. The gee-pole is used to control and help start the sledge. It consists of a stout hardwood pole, 1½ to 2 inches in diameter and 6 to 7 feet in length (fig. 6). This pole is attached at a 45° angle to the fore part of the left runner by lashing the pointed bottom end of the pole to an iron ring on the runner and to a triangular-shaped iron on the front of the sledge. Some types allow for an adjustment in height. In the Nansen-type sledge, the gee-pole is fastened directly to the sledge. The upper end extends

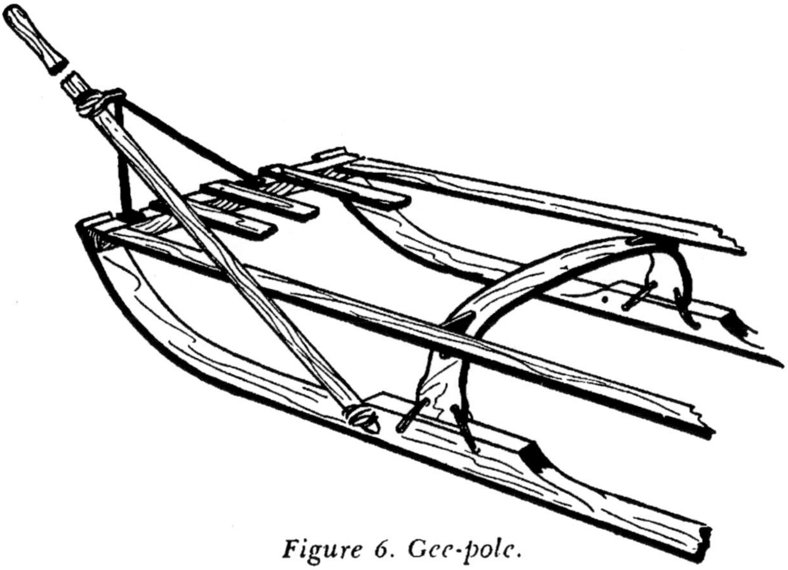

Figure 6. Gee-pole.

forward and upward, so that when a man stands beside the sledge his right hand, as it falls naturally to his side, may reach and grasp the end, affording a simple but effective means of control.

b. Handle bars. Every komatik, basket sledge, or toboggan used in rough country should have handle bars (figs. 4, 30). They are necessary for steering and greatly facilitate control of the sledge. Handle bars should be approximately hip-high. As they must be an integral part of the sledge, they should be lashed to it and made rigid by a cross bar or by two ropes tightly fastened crosswise between the two uprights.

c. Runners. (1) Except in very low temperatures, $\frac{3}{32}$-inch or $\frac{1}{8}$-inch nickel or spring-steel, german silver or duraluminum, shoes or runners should be used over the wooden runners of all types of sledges. The metal shoes should be screwed or bolted to the runners; holes made for this purpose should be countersunk.

(2) In extremely cold weather, the snow surface acts somewhat like sand on the runners, and at such times it is desirable to cover the steel runners with demountable hickory shoes. These shoes should be from 1½ to 4 inches wide, and ½ inch to ¾ inch thick. The width of the shoe will vary in accordance with the conditions of the region in which it is to be used.

① *Hinge brake.*

Figure 7. Mechanical brakes.

d. Mechanical brakes. (1) The mechanical brake, which is used on the basket-type sledge, consists of a 6-inch board running the length of the sledge along the middle undersides of the bridges. Two to five 3-inch iron hooks or claws are bolted to the rear end of the board. The front end and the middle are permanently fastened to the sledge, while the rear end is held up by the spring in the board and a mechanical steel spring attached to the top cross rail of the sledge. The brake is applied simply by stepping on the board to bend it downward, thus plunging the iron hooks into the snow surface. This is done while standing on the rear ends of the projecting runners.

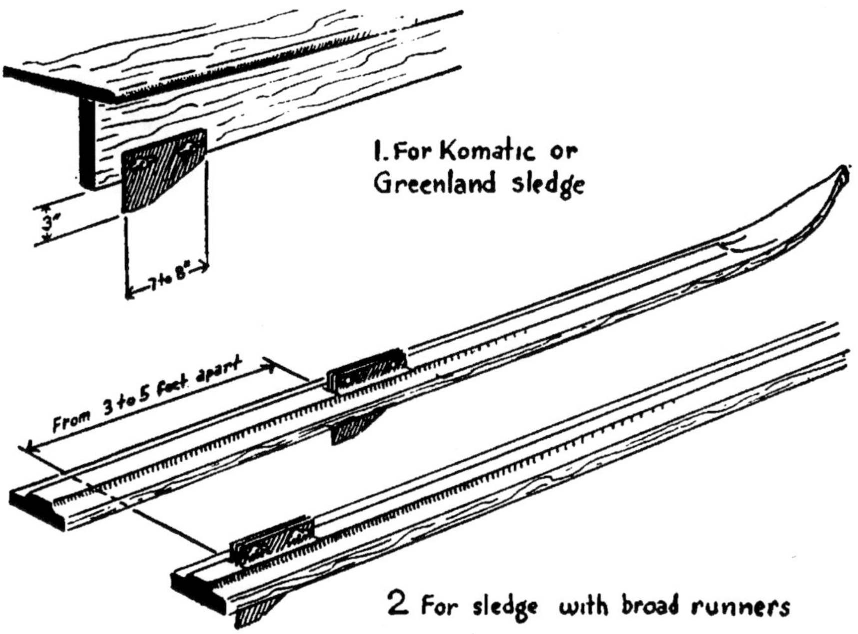

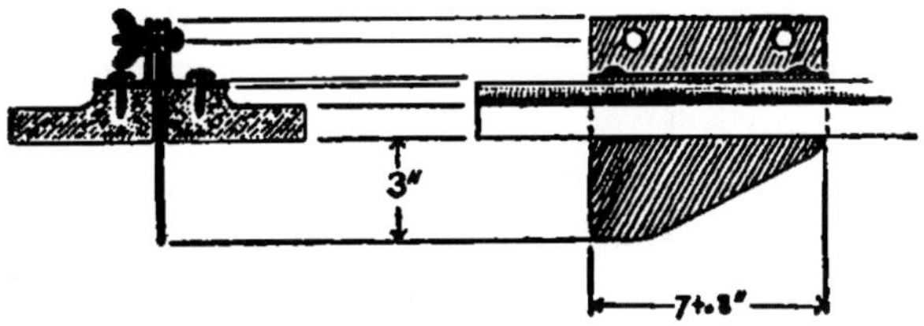

② *Slew brake.*

Figure 7. Mechanical brakes -- Continued.

(2) A modification of this type of brake is the *hinge brake* (fig. 7①), which is more efficient and less likely to

catch on rocks, pieces of ice, or other obstacles over which the sledge is being dragged.

e. Slew brakes. When sledging on hard icy surfaces or traversing slopes, any type of sledge will tend to side-slip. This tendency can be offset to a large extent by the application of slew brakes. These are detachable, flat, triangular pieces of steel which are applied as shown in fig. 7②. The brakes should not be opposite each other. One should be near the back end of one runner, and the other about 3 feet farther forward on the other runner.

f. Rope brake. (1) A simple brake for use on the freight-type sledge may be made of ¾-inch rope. While the majority of rope brakes of the roughlock type require that the sledge be stopped while they are put on, the simple rope brake can be applied from a position at the gee-pole while the sledge is in motion. The rope used is slightly longer than the width of the sledge. Each rope end is spliced in a loop around one of the runners between the front endpiece and the first bridge. (The loops should be large enough to be looped back over the projecting end of each front runner when the brake is not being used. If the end of the runner does not project, a small hook may be screwed into the wood to hold the rope.) To apply the rope brake, slip each loop off the runner or hook; the rope will then slide down the upturned runner and under the sledge. The first bridge will prevent the rope from sliding too far under the sledge (fig. 11).

(2) Roughlock brakes, used with a Greenland komatik, are made of ropes 10 feet long which are spliced in loops. The loops may be thrown over the fore part of the runners without need for first stopping the sledge. As a rule, two loops are adequate. On ice or hard-crusted snow, it is sometimes necessary to use chains for roughlocks; but the use of chains on sledges other than the komatik should be avoided as far as possible, as they are likely to damage the runners.

Section II. HARNESS

10. GENERAL. a. The harness is a very important part of the sledging equipment; if it is ill-fitting or of an improper type, it will injure the dog as well as reduce his pulling efficiency. There are many styles of harness, such as the Siwash, the collar, and the single tandem. Each of these in turn may be either the high or the low draft type. In the high draft type the trace-line comes off the harness from the top of the dog's back; and in the low draft type, it comes off the dog from around the upper part of his hind legs. Each has its individual merits, and the style used will depend upon the terrain and the loads to be pulled.

b. A proper harness should place the pulling strain on the dog's shoulders and chest. In high draft harnesses the tug rope is fastened from the top of the back; a disadvantage of this type is that with heavy loads, in addition to the strain on the dog's shoulders and chest from the forward pull, there is also a downward pull from the tug across the back. This downward pull is likely to be tiring to the dog's hind legs.

c. The following are some of the causes of poor fit in a harness:

(1) The holes for the dog's front legs may be too small and the straps too near the junction of leg and body for comfort, causing friction to wound the dog. The strap passing under the leg should be at least 2 inches away from the pit of the leg.

(2) The collar opening for the animal's head may be too small or uncomfortably far back. When this happens, the pressure of the straps on the dog's throat causes him to gasp. In this case, the point at which the chest straps cross should be lowered, so that it is at least 2 or 3 inches from the base of the throat.

(3) The singletree may be so near the dog's body that it presses against his thighs.

(4) The singletree may be so far from the dog's body that it permits the harness to pass over the body.

d. Materials used for harness construction may be

leather, 1- to $1\frac{1}{2}$-inch webbing, or lampwick. Webbing will prove best, for dogs will not chew it as readily as leather. In addition, it is cheap, strong, easy to sew, and will not become as stiff as leather during cold weather. Leather harness is stronger than webbing but requires oiling and care. Native tanned sealskin or bearskin may be used if necessary. Leather collars are best purchased commercially.

e. Types of harness. (1) *Siwash harness.* (*a*) The low-draft, web Siwash harness (fig. 8) was extensively used

Figure 8. Siwash harness.

on the United States Antarctic Expedition, and is being adopted in the Army sledge-dog training program. In the sledge-dog breeding and training program conducted by the Russian government on Dickson Island in Arctic Siberia, it was found that the Siwash "chest" harness affords greater protection to the dogs and reduces their expenditure of energy. The design of the felt-padded neck yokes places the pulling strain on the shoulders and chest of the dog.

(*b*) It is advisable to make the Siwash harness adjustable to all dog sizes. This can be accomplished by placing a buckle (with a sufficiently long overlap of the webbing to allow for size increase) in each of the rear traces and in each of the cross backstraps, several inches in front of the singletree. Leather or canvas patches should be sewed to the strap to keep the buckles from hurting the dog.

(*c*) The Siwash harness is light, easy to repair, and less likely to be chewed by the dogs than other types. Moreover, it fits better than the collar or tandem-type harness.

(2) *All-purpose collar harness.* This is the same type

as the tandem harness described in (3) below, but has a singletree which permits its use with the double-tandem hitch. Care should be taken to insure that the collar fits properly. A bad fit is indicated by the wheezing of the dog. One should be able to put two fingers easily in the space between the throat of the dog and the bottom of the collar. In using the collar type harness with the double-tandem hitch, see that the outer side trace does not come over the back; if it does, it will place downward strain on the dog's hind legs.

(3) *Tandem harness.* (*a*) This type of harness (fig. 9) is used only with the single tandem hitch. It should be

Figure 9. Tandem harness.

of the collar type and constructed of leather. As the outside traces bear all the strain, they should be made of good, heavy leather. A heavy singletree is necessary for the attachment of the rear harness to the sledge bridle.

(*b*) The collar and the tandem harness are rather heavy and generally do not fit as well as the Siwash harness. Furthermore, since they are made of leather, they are harder to repair and are more likely to be chewed by the dogs.

f. Collars. (1) *Strap collars.* Strap collars should be of good grade leather, 1 inch wide and 20 inches long. A "D" ring for attachment of chain snaps and neck line (double tandem hitch) should be sewed to the collar in the leather overlap which goes around the buckle. A 1-inch "keeper" or leather loop is set between the "D" ring and the buckle.

(2) *Choke collars.* Vicious dogs and dogs that attempt to break their chains may be fastened with a choke collar made of leather-covered cable or chains. An 18-inch size will fit the average dog, but a few 20-inch and 22-inch sizes may be required.

g. Singletrees. The purpose of the singletree is to spread the traces so that there will be no rubbing on the dog's flanks. Singletrees may be made of hickory or any other wood that will not split easily. The dimensions should be 12 inches by 1 inch by ¼ inch, or heavier. A continuous trace which encircles the singletree should be used. The singletree can then be adjusted to the dog's length by shifting it either backward or forward. (If the harness is adjustable this will not be necessary.) The end of each singletree is fastened to the side trace by bending a ¼-inch metal strip into a loop around the trace and screwing the end of each strip into the edge of the wood, or by using a long rivet through the wood. Slipping of the trace is prevented by placing a small screw through the metal strip and trace into the end of the wood. The tug line from the towline is attached with a clove-hitch knot to the loop formed by the trace back of the singletree. This method takes the strain off the ends of the singletree.

h. Tug rope. (1) For use with the double tandem hitch, a ¼-inch or ⅜-inch rope, 30 inches long, is best. The tug ropes (also called traces) for fan hitches are longer than for other hitches. (The true fan requires tugs of identical length, usually about 18 feet long. For the modified fan, the shortest tug should be about 6 feet long.) A loop should be spliced in one end of the tug rope. This is fastened to the harness trace by a clove-hitch knot, and a swivel snap is spliced to the other end. If the double tandem hitch is used, fasten the swivel snap to an iron ring on the towline; with the fan hitch fasten it to the ring of the bridle.

(2) Tug ropes are very frequently chewed in half by some dogs, a practice which is the subject of much hard language on the part of drivers. If the dogs persist in chewing, they can frequently be discouraged by dipping the line in a solution of tar or creosote. If this fails, it may be necessary to use a thin chain, instead of the tug rope, on the harnesses directly in front of the offending dog.

i. Neck line. When the double tandem hitch is employed, a 15-inch overall length of 1/4-inch rope may be used for attaching the dog to the towline. One end is spliced or fastened to the iron ring on the towline section with a clove-hitch knot, while the other end is attached to the "D" ring in the dog's collar by means of a 3/8-inch lineman's swivel snap. However, the neck line as well as the tug rope is in an excellent position for chewing, and much trouble may be experienced. To discourage dogs that have the chewing habit, the line can be creosoted or tarred. If this fails, a 2–0 weldless chain with six and one-half links per foot is recommended; it should have a swivel snap on each end. The primary objection to the use of a steel chain is that it will injure the dog's mouth if he attempts to chew it in extremely cold weather. Good training (par. 48d) may break a dog of the habit of chewing these lines.

j. Towline. (1) The towline (sometimes called gang-line) used in the double tandem hitch consists of one 68-inch (overall length) section for each pair of dogs (fig. 10). Although rawhide or leather is good for this purpose, the dogs will chew it more readily than rope.

(2) The first two rear sections should be of 3/4-inch rope (with a nine-dog team), while the remaining sections can be of 1/2-inch rope. A loop is spliced at each section end, using a tapered splice. A 2 1/2-inch ring made of 3/16-inch iron, for attachment of trace and neck lines, should be spliced into the front loop of each section. This will place a ring at each sectional joint. The first two sections of heavy rope can be permanently spliced together; and the others may be joined by clove-hitch knots. This allows for an easy adjustment in the towline if a

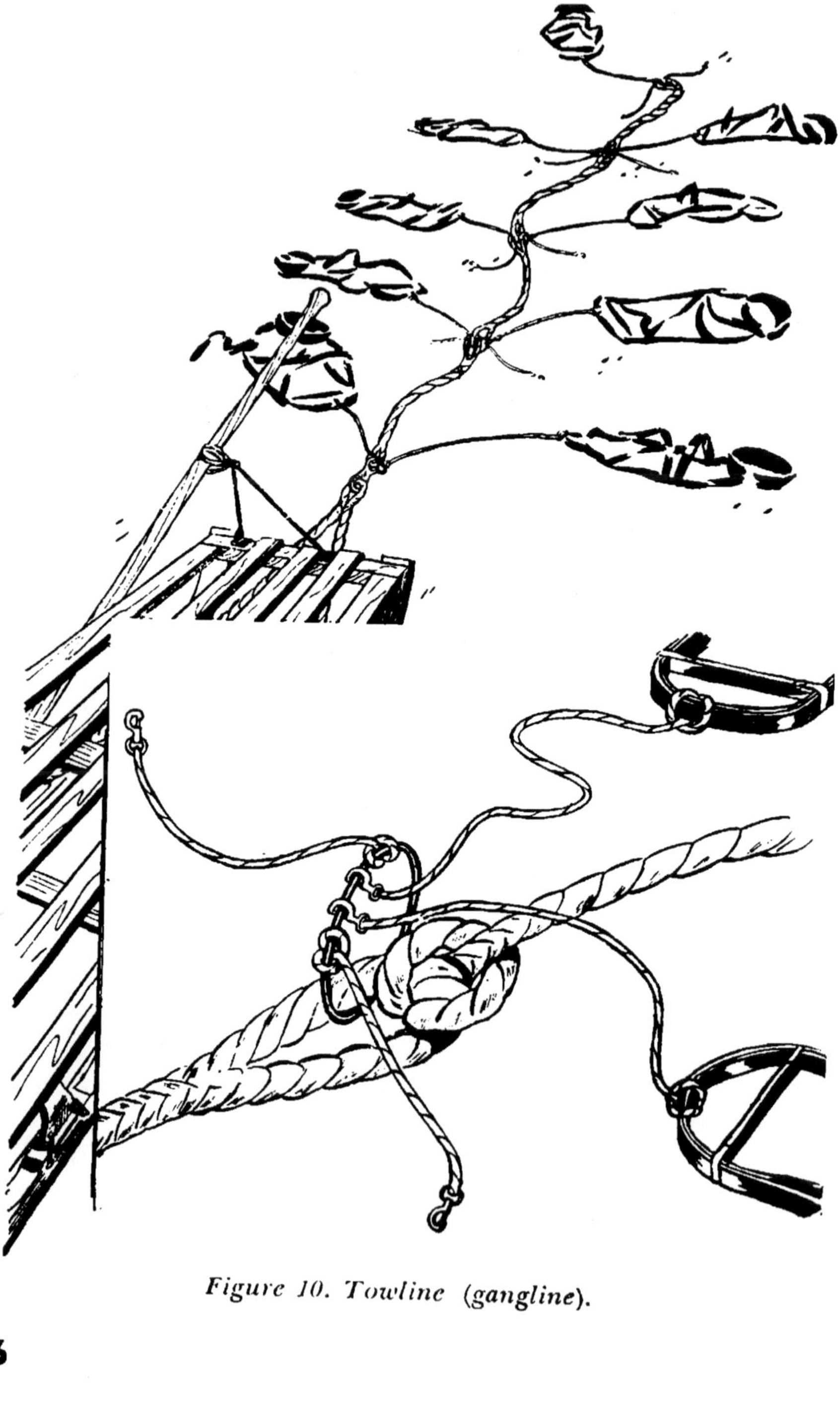

Figure 10. Towline (gangline).

change in the number of dogs is desired. All sections may be permanently spliced if no change in the team is contemplated. If heavy loads are to be hauled over long distances, iron rings should not be used to hold the sections together, as the constant heavy strain of the metal on the rope will quickly wear through the rope. A clymer or lineman's climbing snap, with concealed spring ($1\frac{1}{8}$-inch eyelet), is spliced or lashed to the rear end of the towline. This snaps to the ring of the bridle attached to the front of the sledge.

k. Bridle rope. A $\frac{1}{2}$-inch or $\frac{3}{4}$-inch rope may be used for a bridle rope. The ends of the rope are spliced

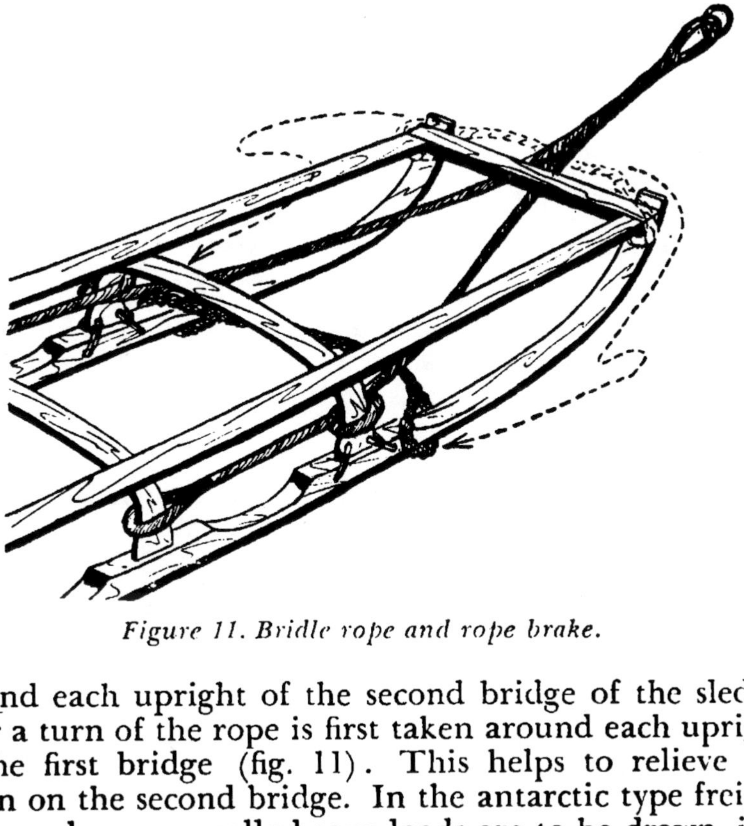

Figure 11. Bridle rope and rope brake.

around each upright of the second bridge of the sledge, after a turn of the rope is first taken around each upright of the first bridge (fig. 11). This helps to relieve the strain on the second bridge. In the antarctic type freight sledge, when unusually heavy loads are to be drawn, it is advisable to splice the end of the bridle (at the second bridge) through a rawhide thong loop at the rear of the second bridge. This loop can be formed by drilling a hole through the upraised width of the runner and passing a thong through it. Fastening the bridle in this manner will put the strain on the bearing surface and relieve strain on the bridge itself. A ½-inch or ¾-inch iron honda, with a 2½-inch steel ring, for attachment of the snap on the end of the gangline, should be lashed into the forward end of the bridle. This end must extend at least 12 to 15 inches ahead of the front end of the sledge and should be suspended from the under side of the sledge by means of a sling or swing loop, to prevent it from dragging on the snow. The bridle may be waterproofed by soaking it in linseed oil or by wrapping it in friction tape. If tape is used, it should be varnished or waxed with a floor wax to prevent peeling.

l. Swivels. The round-eye swivel, with either the bolt-action snap or the concealed-spring snap, is satisfactory. Snow and ice may sometimes become lodged inside the concealed-spring snap and make it difficult to open; this will not happen with the bolt-action snap. The bolt on the bolt-action type may occasionally freeze, but breathing on the frozen part will generally make it work again. The swivels appear to work best with the concealed-spring type. The open-spring snap is not recommended because the spring may sometimes catch in the dog's hair or skin.

Section III. MISCELLANEOUS EQUIPMENT

11. CANVAS SLEDGE TANK. a. This is one of the most important items of equipment for the freight sledge. Tanks have been used on all the Byrd and United States Antarctic Expedition field operations. They are simply canvas "boxes" made of 10-ounce duck (fig. 12), which will keep gear intact should the sledge roll over or fall into a crevasse. Their weight is negligible compared to wooden boxes or large heavy tarpaulins. Tanks make for less shifting of equipment and cut down the danger of loss. They keep equipment dry and materially reduce weight. When traveling through thaw pools, parts of the sledge are often under water; if tanks are used, they help float the sledge and keep the equipment dry, provided the canvas does not become thoroughly soaked.

b. The size of a tank will vary with the sledge. It should be as wide as the sledge itself, as long as the distance between the front and rear bridges, and 2 feet in height. The tank takes on a rectangular shape when

Figure 12. Canvas sledge tank.

filled. The floor, or bottom piece of canvas, should have an overlapping edge to allow for the sewing-in of grommets (flat metal eyelets), if not already stamped in, which are used to lash the tank to the sledge. The grommets should be set in at each corner and in the sides above each bridge. Lashing is made around the bed raves (top slats) and bridges. The overlapping flaps on the top ends and sides of the tank should be at least 3 feet in length.

12. TETHERING LINE OR CHAIN. **a.** Dogs should be kept secured when they are not being driven. On trail operations a 3/16-inch or 1/4-inch steel airplane cable, sufficiently long to hold one team (fig. 13), should be used, allowing 7 feet per dog. Choke chains may be joined together if cable is not available. If cable is used, a loop through which an anchoring stake can be driven is spliced on each end. Each dog is secured to the line by a 2-foot kinkless, twisted machine chain which has a swivel snap at one end for attachment to the dog's collar. This chain is fastened to the line by means of a collar spindle around the line, with the end ring of the short chain welded around this spindle. If this cannot be done, a small ring may be tied in the line and the short chain may be snapped to it (fig. 13). These short chains should be 7 feet apart to prevent the dog from reaching his neighbor.

Figure 13. Tethering line.

b. The use of a tethering line has many advantages: it makes for easier feeding; it eliminates the risk of injury to the dogs through fighting; it prevents harness and other equipment from being damaged by loose dogs. Its worth is appreciated by any dog driver who has been routed out of a warm sleeping bag in the middle of a cold night to separate a pack of fighting dogs.

13. SLEDGE ANCHOR. This type of brake is commonly used in the western Canadian arctic, both for stopping the sledge while driving, and for holding it in place once it has stopped.

a. The Hudson's Bay Company manufactures such an anchor (fig. 14); but the construction is simple and it may be easily improvised. The material required consists of a 10-foot, ½-inch rope, a 2-inch iron ring, a 4-inch iron

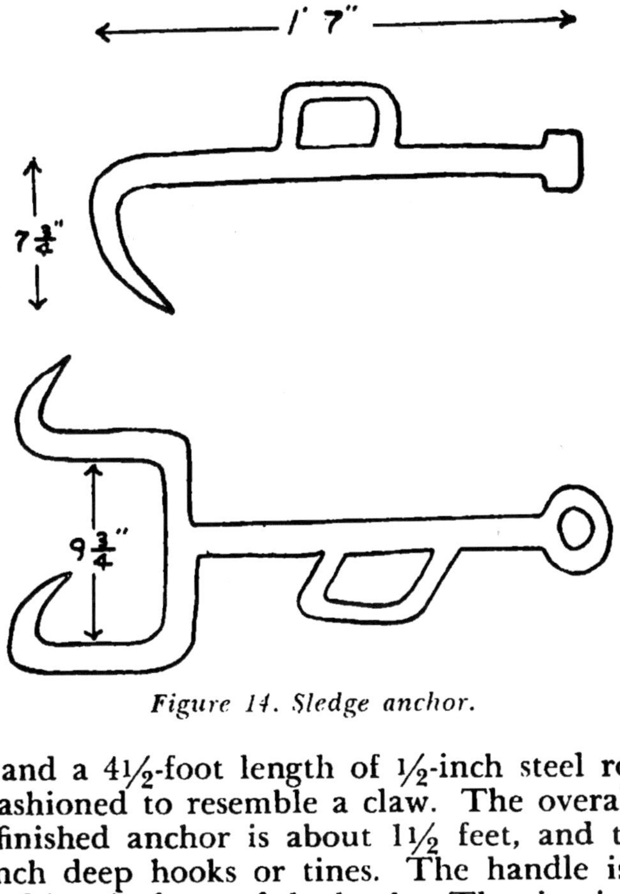

Figure 14. Sledge anchor.

handle and a 4½-foot length of ½-inch steel rod. The rod is fashioned to resemble a claw. The overall length of the finished anchor is about 1½ feet, and there are two 8-inch deep hooks or tines. The handle is welded to the rod just in front of the hooks. The ring is fastened to the fore part of the anchor and one end of the rope is spliced to the ring. The other end is fastened to the bridle or the front of the sledge.

b. Operation of the brake consists of plunging the hooks into the snow at the side of the moving sledge. To anchor the team when stopped, the anchor rope is tightened and the hooks are kicked deeply into the snow.

14. CHOKE CHAIN. **a.** This is a combination choke collar and kennel chain, used mainly to secure the dog at the kennel. It can also be used on the trail as a tethering line by joining several chains together. The choke chain should be at least 6 feet long; the best type is made of a kinkless, twisted machine chain.

b. It is good practice to carry several of these handy chains on the trail, where they can be attached quickly to the sledge for use as a roughlock brake, as a towline, or for holding back the sledge by hand.

15. DOG BOOTS. **a.** Snow and ice lodging between the toes, sharp ice which cuts the pads, and the salt in sea-ice

Figure 15. Eskimo dog boot—open type.

tend to injure a dog's feet and render him useless for pulling. When these conditions are encountered a protective boot must be used.

b. It is difficult to construct a completely satisfactory boot which will stay on. Dogs object to wearing them and will often attempt to chew them off; it is therefore necessary to use either drawstrings or friction tape to hold the boots on securely. Because the strings must be drawn rather tightly, they cut off circulation, and it is necessary to remove the boots upon completion of the day's sledging. Boots can be made of 8- or 10-ounce canvas duck. Green sealskin, if available, is also excellent, but it must be kept moist in order to be form-fitting, and all blubber must be scraped from the skin.

c. Some drivers prefer an open boot which protects the feet and still allows ice particles to drop out. Others prefer a closed boot which does not allow ice to get directly next to the foot. Whatever the type of boot used, it should be simply constructed.

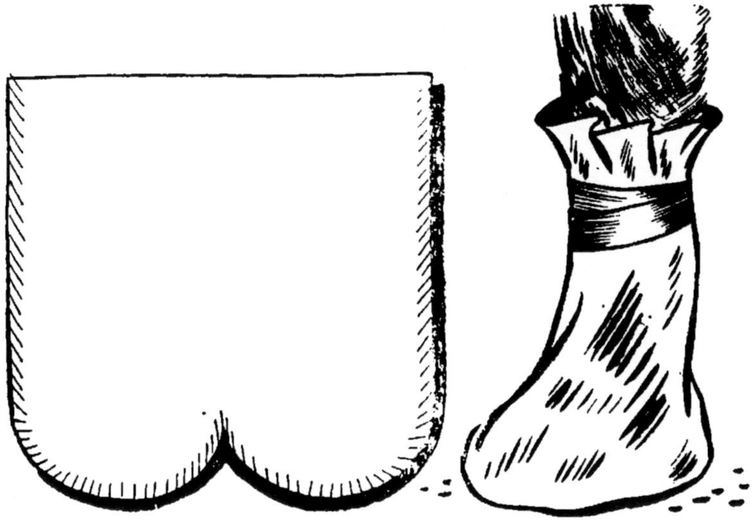

① *Sewed type.*

Figure 16. Closed types of dog boot.

(1) A more or less standard Eskimo boot is an open type of very simple design, which will be found most satisfactory (fig. 15). This type is about 3 inches wide, and is made so that the two front toenails come through the boot and hold it intact on the foot, at the same time permitting a grip on glassy ice surfaces. No sewing is required in this type of boot (which is a great advantage

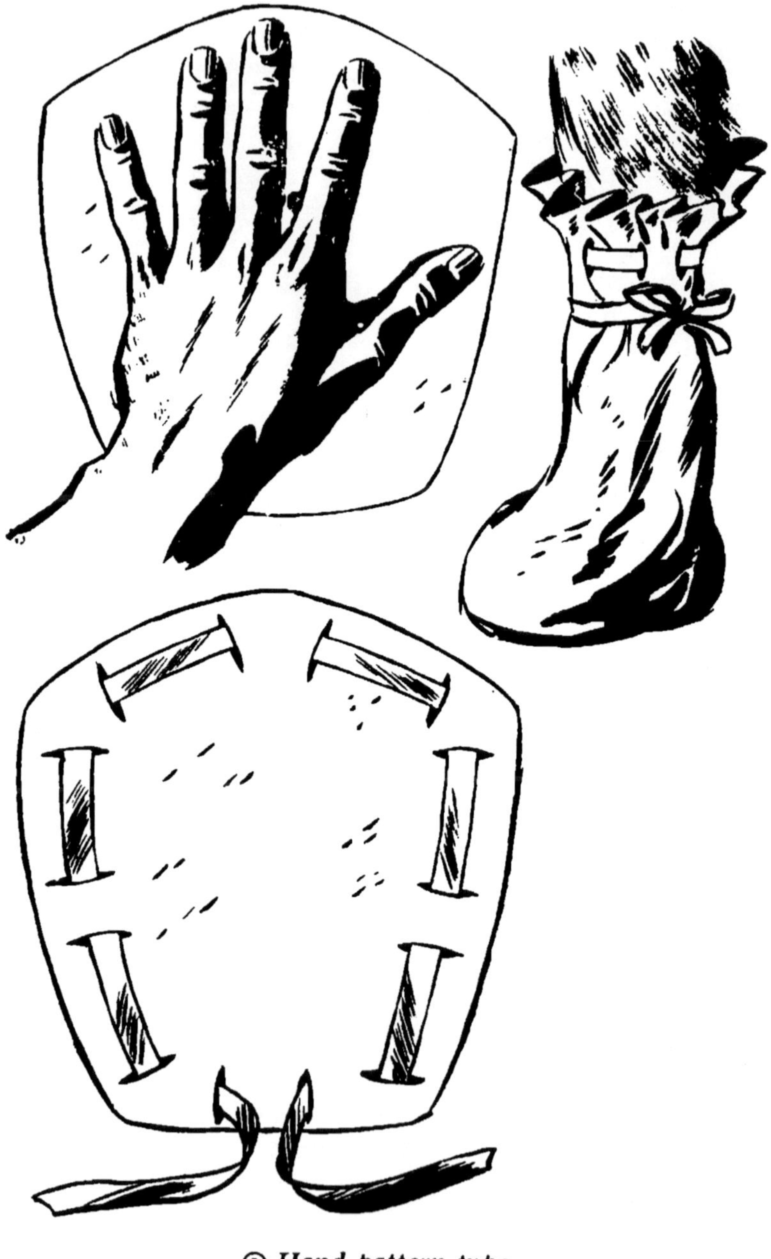

② *Hand pattern type.*
Figure 16. Closed types of dog boot — Continued.

while traveling); thongs looped through one of the folded-up ends tie the boot on the foot. The boot can be turned over when one side is worn out.

(2) The closed type of boot provides more protection for a badly cut foot than the open boot. One type of closed boot requires sewing, (fig. 16①). Another and better type can be made easily on the trail, using the outline of a hand as a pattern (fig. 16②). Whichever type is used, it is essential that the boots be taken off at night.

16. WHIPS. a. Eskimos, when driving dogs in the fan hitch, commonly employ a short whip-handle which has

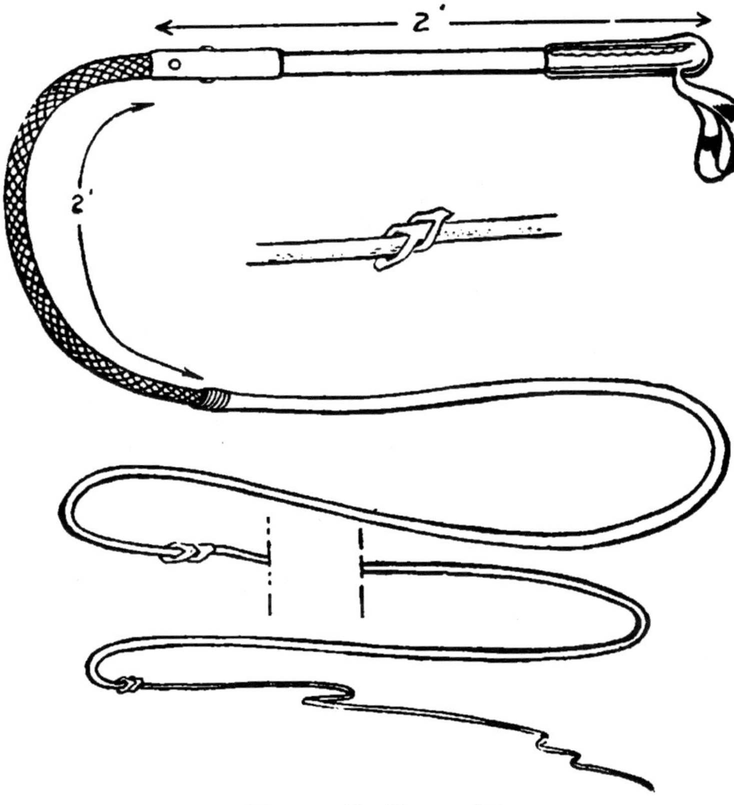

Figure 17. Dog whip.

a thin, 20- to 30-foot sealskin lash attached to the end. A good dog driver knows how to handle such a whip with precision, but uses it only in extreme emergencies. The carrying of a whip by other than experienced drivers should be discouraged. A 6-foot flexible, tapering, leather whip is used most commonly among sledge drivers—mainly in case of a fight among the dogs. This type of whip is the most easily procured.

b. It is easy to make a good, long, dog whip, a above, from an ordinary bull whip. The handle may be made of a broken ski-pole (a piece about 2 feet long) from which snow-disk and point have been removed. The tubular metal piece that protects the joint should be left in place. To insure comfort for the whip-hand, remove the leather grip from the ski-pole and fit it over the top of the improvised handle. Then cut a 2-foot length from the bull whip so that the thickest end of the cut part can be fitted into the tubular end of the handle. A couple of nails jammed through two holes bored in the tube will hold handle and base of whip together. A thong of ½-inch saddle leather, from 12 to 15 feet long, is then sewed to the end of the bull whip. Different sections can be assembled as shown in figure 17. The thong should be rounded and graded from ½ inch to ¼ inch, or less. To complete the whip, attach a 3-foot leather lace to the final section of thong. Grease the whip and keep it greased to insure continued suppleness.

17. TRAILER HITCH. The hitch connecting the front and trailer sledges consists of two ½-inch ropes, 11 feet in length. If the trailer sledge is to be permanently used, one end of each rope should be spliced around each front bridge upright of the trailer sledge, or if loads are heavy, in the manner described in paragraph 10k. A loop should be spliced in each forward end. Hitching consists of crossing the ropes (the rope around the left bridge on the trailer sledge goes to the right rear bridge of the front sledge, and *vice versa*) and bringing the looped end around the rear bridge upright of the front sledge. Securing is done by means of a wood toggle (fig. 18). The trailer sledge should travel in the track of the front sledge. Adjustments for the length of each rope can be made with the toggle. In deep snow it may be necessary to raise the two ropes by means of a sling under the front piece of the trailer sledge.

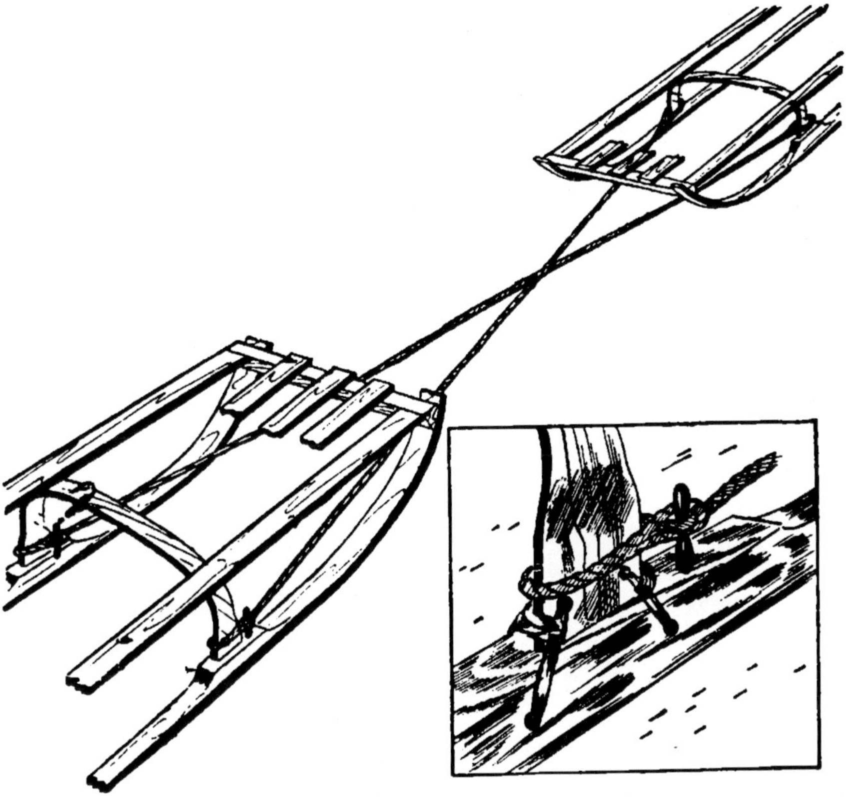

Figure 18. Trailer hitch and toggle.

18. "OUIJA BOARD". a. The "ouija board" is similar to a short toboggan. Usually it measures 3 feet in length and about 1½ feet in width. It can be constructed easily from two or three boards, cleated together at each end and at the start of the turn. The upturn is formed by steaming or soaking the ends and then bending them. The ouija board is then equipped with a short bridle. An extra section of towline is placed immediately in front of the sledge, and the bridle rope of the ouija board is snapped to the ring at the front end of this section. This places the board under the towline and directly between the front of the sledge and the rear of the dog or dogs (fig. 22).

b. The ouija board is used in certain sections of Alaska and the Yukon Territory. It is employed with the single or double tandem hitch in brushy or wooded regions where trails are narrow. The driver merely stands on the board with the towline between his legs, holding on to the gee-pole, which, when using the ouija board, is on the right side of the front of the sledge. Some Indians use short skis in place of the ouija board.

CHAPTER 3

KENNEL MANAGEMENT

Section I. PERSONNEL, FACILITIES, AND ADMINISTRATION

19. PERSONNEL. a. An officer should be placed in charge of the management of the kennel. It will be his duty to instruct his men in the care and handling of the dogs and to supervise the prescribed kennel routine and management procedure. A noncommissioned officer should supervise and coordinate the training and the work done in the kennels. One private or private, first class, can take care of up to 15 dogs. Best results in the feeding, care, training, and exercising of one team (usually consisting of 9 dogs) are attained when the same driver is assigned to and handles the same dogs over a long period of time. The ideal for the Army is to accustom a team to two men and two men only. Only one of these men, however, should handle the dogs at a given time.

b. Selection of Personnel. Considerable care should be exercised in selecting men to work with dogs. First of all, a dog handler must like dogs; if he lacks interest or enthusiasm, any other assets he may possess will prove useless. He must be patient, firm, intelligent, and fairly athletic. A sense of humor and an even temper are assets; there will be times when both will be severely taxed. Cruelty has no place in dog training or driving, and a handler who displays cruelty should be relieved of his job. It is well to consider also the disadvantages of sledging personnel who wear glasses; glasses are subject to loss or breakage on trail operations, and their continual frosting will prove a serious handicap.

20. KENNEL LOCATION. Sledge dogs are notorious for their howling. In a camp where this may be disturbing, it is well to locate the kennels at a suitable distance from the post. It is important to select a well-drained and sandy, or gravelly, area. Rocky ground is hard on the dogs' feet and makes the securing of tethering lines difficult. The prevailing winds should be determined and the kennel placed to leeward of the camp. The kennel openings should also be placed to leeward of the prevailing winds. Failure to observe these rules will result in continual filling of the kennels with drifting snow as well as in discomfort to the dogs.

21. KENNEL CONSTRUCTION. a. Construction of an elaborate kennel is usually impractical at an arctic outpost. The most effective type of kennel is an individual crate or "house" for each dog (fig. 19). This is designed

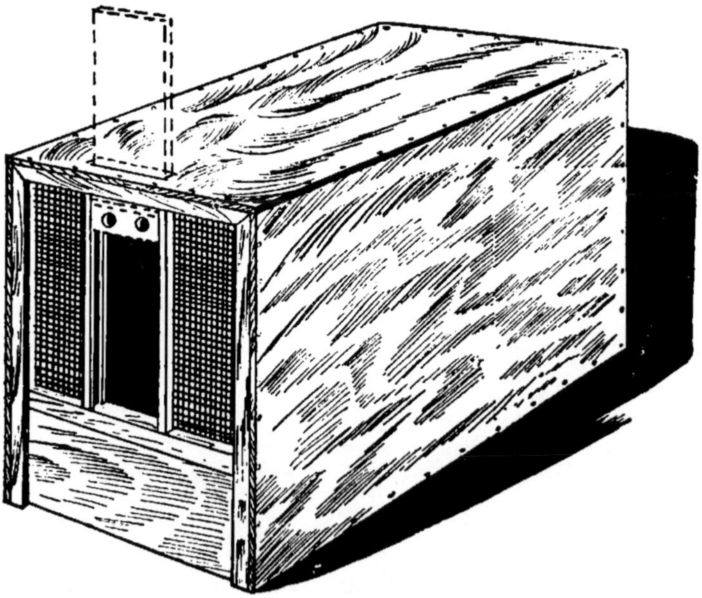

Figure 19. Shipping-crate kennel.

as a combination shipping crate and kennel. The kennels should be located as suggested in paragraph 20.

b. In cold, windy, and snowy areas, the dogs should be placed in a protective tunnel, with each dog in its own kennel. If the snow is sufficiently hard and deep, a tunnel can be cut under the snow.

c. If this is not feasible, the individual kennels should be placed in a straight line and covered with a tunnel of well-braced canvas or board. One team should be placed in each tunnel. If the tunnels are placed parallel, the drifting snow will soon fill the intervening spaces and provide good protection for the dogs. Keeping the dogs in such a tunnel arrangement will also eliminate the possibility of the kennels being filled by drifting snow. This spares the dogs considerable discomfort and saves much work for the driver.

22. KEEPING RECORDS. Records are important in the management of a well-operated kennel. The following data should be kept:

a. Individual record. This record includes name, serial number, dam, sire, age, sex, breed, markings, weights, inoculations, and worming dates. If the animal is a bitch, the dates of mating, number in litters, and other pertinent data should be recorded. The dog's name and serial number should be stamped on its collar. (See W.D., A.G.O. Form No. 10–24; W.D., Q.M.C. Form No. 120 may be used until exhausted.)

b. As soon as acquired, dogs are tattoo-branded on the inside of the left flank, using one letter and three numerals. The brand numbers will be assigned by The Quartermaster General.

c. Record of kennel property. (1) A food record must be maintained at the kennels in order to insure the correct expenditure and requisitioning of foodstuffs.

(2) The quantity of food fed each dog per day is in proportion to the amount of work done. After the training program is fixed, the amount of food required for a certain period is computed by the simple process of multiplying the pounds per day per dog by the number of days in the given period. It is necessary to estimate in advance the number of dogs that will be on hand, taking

into consideration the expected number of grown dogs and the estimated number of puppies born.

(3) After the food has been divided into portions for different periods, a constant check is necessary to keep the consumption of food within the budget allowance. A food shortage is likely to occur if the stock is carelessly expended.

Section II. FEEDING AND CARE

23. GENERAL. a. The feeding of dogs will vary greatly with the region, season, and type of work carried on. All bases should be amply supplied with a good, all-purpose, commercial dog food. Meat and fish, however, will keep the dogs in better condition, and these foods should be incorporated in the diet whenever available. They may be purchased from local natives or they can be obtained through hunting and fishing by the base personnel. In cold weather, when fresh food is plentiful at the base, the dogs should have at least 3 pounds of food daily. They should be fed as scientifically as possible and trained to take foods which they may have to eat on the trail. It is advisable to train dogs to eat a wide variety of foods while they are young and during the training period. This simplifies the problem of procuring food on the trail or on transfer.

b. Walrus or seal meat is regarded as the very best food for dogs. White whale, whale, narwhale, polar bear, reindeer, caribou, and fish also are good. Salmon is most commonly used in Alaska, while cod, hellefisk, halibut, salmon-trout or char, and shark are the most nutritious fish in the eastern Canadian arctic and Greenland areas. Fresh shark flesh is poisonous and should never be fed to dogs except in food emergencies, and then only in very small amounts. If boiled for several hours or dried, or after it is a week old, shark flesh can be eaten safely and is fairly good dog food.

c. Oatmeal, corn meal, and rice, commonly used in Alaska, are good though rather starchy. Whenever possible, use all the better table scraps from the base kitchen. However, small bones such as chicken bones should be avoided.

24. PREPARATION AND PROPORTION OF FOOD. a. Commercial dog foods, if mixed with water or meat broth, should be prepared at least an hour before feeding to allow time for absorption of moisture and swelling of food. The amount fed will vary with the exercise, condition, and size of the dog, but generally 2 to 3 pounds will be adequate for a 70- to 80-pound animal. One or 2 pounds of dried fish, or 2 or 3 pounds of fresh meat, per meal will keep a dog in excellent working condition. Twice a week, cod-liver oil should be added to the dog food, if available. A rule often followed in feeding is to give an idle dog half as much food as a working dog. However, this depends again on the condition of the dog. Observe each dog carefully to see whether all the food given him is eaten; if not, the amount should be reduced.

b. Some dogs will occasionally refuse to eat when a new type of meat is fed to them. If it is consistently refused, the new meat should be set aside until it becomes partly putrified. The smell of the new meat is lost in the putrefaction, and the dog will then generally eat the meat. Apparently all aged meat has much the same smell, and a dog cannot differentiate between the different kinds. Aged meat is excellent dog food.

25. WATERING. Usually it is not necessary to water dogs if clean snow is available. If there is no snow, the dogs should be watered early in the morning and shortly before their afternoon feeding. If dry food is used, it is well to offer water to the dogs at least half an hour after the meal.

26. FEEDING TIME. Working dogs should be fed only once a day, and the feeding should take place at the end of the day's work. When fed before or in the middle of the workday, the dogs often refuse to pull and frequently may vomit. Dogs should be rested at least half an hour after work and before being fed. At the kennel, when the dogs are not working, they should be fed at about 16:00. When in the kennel, a dog in poor condition should be fed twice a day to be brought back into good shape.

27. WEANING OF PUPS. a. At the end of 3 weeks of nursing, the pups should be fed a small amount of cow's milk, or diluted, evaporated or condensed milk. Feed the pups three times daily to augment the dam's milk. Always use the same kind of milk. The temperature of the milk should be uniform (preferably between 55° and 65° F., although this is not compulsory). Raw scraped beef may be added to the diet the fourth week, and finally chopped meat can be fed at the fifth week. Four weeks after birth, the dam should be taken away from the pups during the day and returned to them at night for nursing and sleeping. Any change in food should be made very gradually. Complete weaning of the litter should be accomplished after 6 weeks.

b. Feeding the dam. The dam's food will have to be increased for a short time before the birth of the pups as well as during the nursing period. During the first 4 weeks of the nursing period, the dam should be fed fresh milk or two cans of canned milk daily. The drying-out process starts the fifth week, and after this only one can of milk should be given daily. No milk is fed the sixth week. During the nursing period, at least 50 percent of the dam's diet should consist of meat.

28. FOOD FOR PUPS. The feeding and care of pups has an important bearing on their future development. At the end of the 6-week weaning period, the pups should be put on a diet of oatmeal, canned milk, and chopped meat. The diet can be increased gradually to take in most of the better table scraps available from the mess kitchen. The pups should not be fed small bones which may lodge in the throat or intestines. Potatoes should be avoided. Many dog drivers and breeders have found that pups transferred to a varied diet after weaning show better development in size and strength than pups that are transferred to a strictly meat diet. The pups should be given four meals daily between the sixth and tenth weeks. At the end of 4 months, the number of meals should be gradually tapered to two a day.

29. TRAIL FOOD. a. Trail food must be prepared in advance, although in some parts of the western and eastern Canadian arctic regions it will be possible for experienced hunters to obtain seal or walrus meat while traveling. If game is available on the trail, it will materially reduce the weight to be pulled by the dogs, as it will lessen the amount of food to be carried.

b. Dried fish is a common trail food. It constituted a large part of the dog food used by Amundsen on his trek to the South Pole. Fish can be purchased or freshly caught. To prepare them for trail operations, they should be cut into strips and hung up to dry.

c. If fuel is abundant, cooked rice and fish may be used as a trail food. It has the advantage of taking up very little space. For each dog 2½ pounds of cooked rice (about 6 ounces, dry) are mixed and cooked with ⅛ to ¼ pound of fish. In addition, each dog is fed ½ pound of dried fish.

d. An excellent dog pemmican commonly used on trail operations in the Antarctic is a mixture of a commercial dog food with seal oil or preferably cod-liver oil. Commercial dog food alone will not keep the dogs in good condition over long periods; the high caloric value of the added oil will provide much additional energy. If cod-liver oil is not available, other oil can be easily rendered from seal or walrus blubber. However, lard or bacon grease from the Army kitchen may be substituted. The common method of preparation is to mix the oil and the commercial food into a pastelike consistency, mold the pemmican thus formed into 1½-pound cakes, and then freeze. This makes a convenient, nutritious, compact, and easy-to-pack food—probably the best all-around trail food. If the weather is not sufficiently cold to keep the cakes frozen, the mixture may be stored in 5-gallon tins of the type commonly used in packing dehydrated food.

30. SECURING DOGS. Even though the dogs appear to be friendly among themselves, they must be secured. The tethering line or chain is discussed in paragraph 12. Failure to secure dogs may result in loss of dogs from fighting. Such losses may prove serious when dogs are badly needed.

Section III. BREEDING

31. SELECTIVE BREEDING. a. General. The aim of any sledge dog-breeding program should be improvement of the strain. Only bitches and studs which conform to the standards for the recognized sledge-dog breeds should be used. (See pars. 4–7.) Whenever possible, the records of the grandparents should be checked, for the present bitch and stud may be throwbacks. The condition of the breeding dogs, their past performances and history, and the type of work to be performed by the offspring, should be considered.

b. Standards. (1) The most satisfactory type of dog for Arctic operations has a large chest, moderately long legs, large hairy feet, and heavy fur. (See fig. 1). The male should weigh between 70 and 85 pounds. Breeding should be planned with this ideal in mind.

(2) Heavier dogs are used but they are slower; they require more food; they break through snow-crusts more easily; in general, they do not appear to stand up as well as medium-size dogs under continuous hard work and fast prolonged travel.

(3) Poorly furred dogs require more food than heavily furred dogs to keep warm.

c. Breeding period. A bitch should not be bred before her second period of heat and preferably not before the age of two. The stud dog should be mature. One litter of pups a year should be the maximum. Inbreeding is not recommended. To insure pregnancy, the bitch should be bred first on the seventh or eighth day of the heat period. A second mating should be made after an interval of one day, and if the period of heat is not over, a third mating can be made on the twelfth or thirteenth day. If possible, breeding should be timed so that the pups will be born in the spring. Their training can then begin in the fall.

32. CASTRATION. Castration is not recommended, except possibly in the case of vicious dogs.

33. SPAYING. Spaying is advisable for dogs of poor breeding caliber. Bitches in season have less pulling efficiency and can entirely disrupt a team on the trail. Spayed bitches are usually good workers and their presence will result in less fighting within the team. Some spayed bitches have a tendency to become fat, but if the operation is performed between the fourth and eighth month (before the first period of heat), this condition will not usually result.

Section IV. PREVENTION AND TREATMENT OF DISEASE AND INJURY

34. GENERAL. The treatment of disease and injuries among animals, as well as the prevention of disease, is the function of the Veterinary Corps. However, at Arctic outposts this service may not be immediately available. Thus, it becomes the duty of each soldier to be vigilant and intelligent in the application of simple measures to prevent the occurrence of diseases and injuries among the dogs. Sledge dogs normally are very hardy, but injuries are quite common and diseases often develop which require emergency treatment. In this section, preventive measures and first-aid treatments are briefly discussed. The treatments suggested are intended for use by anyone familiar with the handling of sledge dogs. However, whenever possible, sick or injured animals should be placed in the care of a veterinarian.

35. THE HEALTHY DOG. Health is the condition in which the body functions normally. It is particularly essential that the normal functions of the body of the animal be thoroughly understood; otherwise, a departure from the normal is difficult to recognize.

a. Normal appearance. In the standing position, the four feet are placed squarely on the ground, or with one hind foot slightly in front of the other. The head is alert, eyes wide open, nose moist and free from discharge, and, in prick-eared dogs, the ears are erect. The coat is lustrous and the skin supple and easily moved about over the structure underneath. The color of the mucous membranes of the eyes, nostrils, and mouth is salmon-pink. The bowels are evacuated on an average of three to four times in the course of 24 hours, and the stool should be formed and free from mucus. The color of the stool varies according to the nature of the food. The urine is voided at frequent intervals, and during the course of 24 hours, large dogs will secrete 0.5 liter to 2 liters of urine, varying with the food, the external temperature, and the season of the year. The color will vary from light to dark yellow, depending on the diet.

b. Normal temperature. The normal body temperature of the dog at rest is 101° F., but it may vary 1° in either direction. Immediately after work, and especially in a hot sun, the temperature of healthy animals may rise, but as a rule temperature is unaffected by climatic conditions and is thus a most valuable guide. The temperature is taken with a clinical thermometer in the rectum. The thermometer is moistened or oiled, the mercury shaken down to 96° F., or below, and the bulb of the thermometer is inserted in the anus and allowed to remain for 3 minutes. Then it is withdrawn, the temperature read, the thermometer washed in cold water and the mercury shaken down again below normal. A rise of temperature often precedes any visible symptoms of contagious diseases and is therefore important in detecting infected animals.

c. Breathing. The breathing may be most conveniently noted by the rise and fall of the flanks. The average number of respiratory movements while at rest normally varies from 8 to 16 per minute. These should be counted when the animal is first approached, as excitement or movement causes an increase in the number of respirations. Immediately following exercise, the respiration rate may increase to 60 to 90 per minute. During warm weather, while the dog is at rest, the respirations are greatly accelerated because the skin glands are not active enough to assist in the respiratory functions. In order to overcome this condition, the dog breathes through the mouth and the frequency of respiration is increased.

d. Pulse. The normal pulse rate is from 70 to 100 beats per minute, depending on the breed of the dog. The pulse may be felt on the left side of the chest just in back of the elbow joint, or by placing the fingertips lightly on the femoral artery which is located on the inside of either hind leg. The beats are counted for 30 seconds and multiplied by two. The animal should be at perfect rest when the pulse is taken, as exercise and excitement quicken it. In illness, the pulse is usually faster than normal and its character varies considerably. It should be noted whether the pulse is stronger or weaker than normal.

e. Mucous membranes of eyes, nose, and mouth. Normally, these membranes are a salmon-pink color and free from congested areas and discharges, but during illness they may become red (congested), pale or white (from loss of blood), or yellow (from digestive disturbances).

f. Skin. In health the skin should be supple and roll easily on the underlying structure. The hair should be lustrous, soft, and lie evenly. In disease, the skin is dry, as if adhering to the muscles below; the coat is dull in appearance, harsh to the touch, or "staring," that is, with the hair on end instead of lying flat.

36. INDICATIONS OF DISEASE. Every disease has different indications, and the symptoms vary so greatly that only exhaustive study can acquaint one with them. In order that sick animals may be detected in the early stages of disease, the soldier must be prepared to recognize the more common early variations from normal. The most common preliminary indications of disease are partial or complete loss of appetite, elevation of temperature, accelerated breathing, increased pulse rate, listlessness, a dejected appearance, nasal discharge, persistent coughing, diarrhea, constipation, an unhealthy coat of hair, and unnatural heat or swelling of any part of the body. One of the first and most important symptoms of sickness is impairment of appetite. The best times to inspect animals for evidences of sickness or injury are while they are being fed and during grooming.

37. CLASSES OF DISEASES. Diseases are divided into two classes: communicable and noncommunicable.

a. Communicable diseases. Communicable diseases are diseases that are transmitted from animal to animal, either by direct contact between the sick and well, or indirectly, through the medium of infected kennels, unsanitary feeding dishes, flies, lice, ticks, mosquitoes, or faulty sanitation. The diseased animals throw off in the discharges from the respiratory, digestive and urinary systems the material which will cause disease in susceptible animals. These diseases deserve more attention than the noncommunicable diseases affecting sledge dogs, because they are most likely to appear when dogs are congregated in considerable numbers. Some of these diseases are incurable; others may be transmitted to man, and all may cause great harm if not held in check. Communicable diseases are always marked by a period of incubation, which is the time interval that elapses between infection and the appearance of the symptoms of the disease. This period may vary from a few days to several weeks. Examples of contagious diseases are distemper, rabies, mange, and ringworm.

b. Noncommunicable diseases. Noncommunicable

diseases include all diseases that are not transmissible, either directly or indirectly, from one animal to another. Many of these diseases are caused directly by improper methods of animal management. While noncommunicable diseases are the cause of many lost days of animal service and many animal losses, they do not present the serious problems encountered in communicable diseases. Examples of noncommunicable diseases are nephritis, epilepsy, and nutritional and digestive disturbances.

38. COMMUNICABLE DISEASES—PREVENTION AND CONTROL. a. General health. If animals are maintained in good condition, fed well, kept in clean surroundings, not overworked, and protected from undue exposure, they will resist many forms of infection. Animals in a run-down condition are very susceptible to disease.

b. Segregation. Animals should be divided into three groups: affected, suspected, and healthy. The suspected animals include all those which have been in contact with the diseased. Attendants, watering and feeding arrangements, and all equipment involved in the care of suspected animals should be kept separate until the outbreak is over. Once an animal is placed in the affected group it should remain there until all danger is over. To avoid possible introduction of communicable diseases, all animals are quarantined for at least 21 days upon arrival at any new station.

c. Disinfection. Strict cleanliness of kennels, equipment, water, and feed pans is a good safeguard against the spread of disease. Kennels and equipment which have been in contact with diseased animals should be disinfected with cresolis solution. Articles which can be boiled may be thoroughly disinfected by boiling.

d. Destruction of dogs. Certain diseases are incurable and make it necessary to destroy the affected animal. A dog may be destroyed by shooting him in the middle of the forehead about 1 inch above the level of the eyes.

e. Disposal of dead. The carcasses of all dogs that die as a result of communicable diseases should be burned if practicable; otherwise, they should be buried deeply and covered with quicklime, if available. Ground where discharges from the dead have fallen, should be disinfected.

39. TYPES OF COMMUNICABLE DISEASE. a. Catarrhal pneumonia (broncho-pneumonia). This is an infectious process involving scattered areas of lung tissue.

(1) *Cause.* Pneumonia is caused by certain germs (principally the pneumococcus, associated with other germs). Predisposing factors are chilling of the body surface through prolonged exposure to damp, cold air, associated with fatigue; other diseases which lower the animal's general resistance; and inhalation and aspiration of infectious material in the mouth and nose.

(2) *Symptoms.* The symptoms of pneumonia develop rapidly, usually beginning with a chill. The dog appears depressed and restless. The temperature is high (104° to 106° F.) and shortness of breath is marked; breathing is short, shallow, and very rapid. A short, painful cough is noticed, and the pulse is accelerated (180 to 200 per minute).

(3) *Prognosis.* Unfavorable.

(4) *Treatment.* The dog should be segregated in a heated inclosure. His strength should be sustained by appetizing and nourishing foods such as milk, beef extract, soups and eggnogs given warm and in small quantities every few hours. During convalescence, small amounts of lean meat and other nutritious foods should be given three to four times a day. At the onset, a purgative (calomel 0.03 gram and sodium bicarbonate 0.5 gram) should be given to produce free purgation and stimulate the activity of the kidneys. Expectorants are useful as an aid in expelling the discharge from the bronchial tubes. Applications of counterirritants to the wall of the chest, in the form of mustard plasters, are helpful. The heart action should be stimulated in severe cases by the use of digitalis.

b. Distemper. Distemper is an acute, infectious, communicable disease which in most cases affects young dogs.

(1) *Occurrence.* This is one of the most common dis-

eases affecting dogs and is known wherever dogs are found. It is particularly a disease of young animals. The majority of dogs contract it at some time during the first year of their lives. Whole litters of pups, or all the animals in the kennels, may become affected at one time. Distemper occurs in districts as a local disease. It is most prevalent during the fall and winter months. Highly bred animals are more commonly affected than those bred by natural selection.

(2) *Cause.* The disease is thought to be produced by a living organism which is small enough to escape all filters. Infection may be either direct, caused by animals coming in direct contact with each other, or indirect, through intermediary agents, the virus being taken into the digestive tract with food or drink. A number of factors tend to favor the development of the virus, such as reduction of the general resistance of the animal by poor food, insufficient food, colds, injuries, operations, and various diseases or conditions interfering with the assimilation of food. Pups with weak constitutions are especially susceptible. Ordinarily the disease affects animals from 3 months to 1 year of age. Older animals are very seldom affected and, if at all, take the disease in a mild form.

(3) *Symptoms.* The acute form begins with a moderately high temperature (103° to 106° F.) which may remain elevated for several days to several weeks. The dog is depressed, the hair coat becomes rough and loses its gloss, the nose is found to be hot and dry, and the animal may sneeze considerably. Inflammation of the membranes covering the eyeballs and the inside of the lids is a common symptom. In the early stages there is a serous or watery discharge from the eyes, which later becomes pus. In severe cases corneal ulcers occur, resulting in white scars. Vomiting is an early symptom in a large number of cases. The animal is thirsty and drinks large quantities of water. Constipation is invariably the rule during the early stages, followed by diarrhea in the course of a day or two. The stools are very frothy, fetid, often shiny and sometimes bloody. Early there is a nasal

discharge, first clear, later becoming pus. The dog shows much distress and makes frequent attempts to clear the nasal passages by sudden expiratory efforts,. by rubbing the nose against objects and by clawing at the nasal openings with the forepaws. In severe cases the discharge is so copious that the nostrils may be partially or completely closed. Inflammation of the bronchi is very common, producing severe coughing and labored breathing. The fatal cases have a terminal pneumonia. Nervous symptoms such as excitement, restlessness, a yelping cry, and in certain cases fits, frequently occur. In extreme cases paralysis may follow the disease.

(4) *Prognosis.* The prognosis is not favorable even in the milder forms of the disease.

(5) *Treatment.* The dog should be fed easily digested, nourishing food such as raw meat, finely chopped or scraped, and beef broth, warm milk, or eggnog. The dog should be segregated from other dogs in a dry, warm, well-ventilated room, free from cold draughts. Plentiful fresh water should be provided. During the early stage, calomel (0.075 to 0.1 gram) is used as a purgative. To correct the constipation, warm laxative foods should be given. Eyes should be bathed twice daily with a 3 percent solution of boric acid, followed by a few drops of cod-liver oil in each eye. The nostrils should be cleansed twice daily with a warm solution of boric acid, followed by a few drops of 5 percent solution of argyrol, along with a few drops of glycerine, in each nostril. Distemper serum is highly recommended.

(6) *Prophylaxis.* All pups should be immunized at 4 months of age, using vaccine which is virus of canine distemper. All newly received dogs under 2 years of age should be immunized as above unless a definite record of previous vaccination is shown.

c. Piblokto. This has been a serious disease among sledge dogs in Arctic areas, and it may be well to become familiar with the disease if Army dogs are to be used in these areas. In 1931 there was a rather widespread piblokto epidemic among sledge dogs throughout the eastern Canadian arctic, and in 1936 the disease killed over 80 percent of the dogs at Repulse Bay. The disease is characterized by erratic behavior, foaming at the mouth, fits, and peculiar locomotion. In appearance it is very like the encephalitis disease which affects the white Arctic fox. Some scientists think the fox may contract encephalitis from the lemming which is its principal food. Both the fox and lemming are "cyclic" game species, that is, they reach a high peak in population every 4 or 5 years. It is usually during these high peaks that encephalitis epidemics occur. It is barely possible that Arctic sledge dogs may contract piblokto from contact with infected animals or from eating the carcasses of animals that have died from the disease. The Eskimo habit of turning dogs loose to shift for themselves in the summer causes the animals to rely to a considerable extent on lemmings for food, and this may result in their infection. Feeding on fox carcasses in winter may also be a cause of infection. Therefore, as a precautionary measure, sledge dogs should not be permitted to associate with stray native dogs or to feed on the carcasses of dead animals.

d. Rabies. Rabies is an acute infectious disease, fatal in the majority of cases, and characterized clinically by disturbance of consciousness, marked irritability, and later by symptoms of paralysis. At the present time it is unknown in Alaska.

(1) *Cause.* Rabies is produced by an extremely minute germ and is induced in practically all cases by a bite from a rabid animal, which introduces the virulent saliva into the wound.

(2) *Types and symptoms.* There are two types of rabies: furious rabies and dumb rabies. Symptoms differ depending on whether the brain or the spinal cord is primarily involved.

(*a*) *Furious rabies.* In furious rabies there are three stages of development:

1. Melancholic or prodromal stage. From 3 to 7 weeks after the dog has been bitten by a rabid animal it becomes sullen, afraid, easily irritated, restless, and obstinate, and seeks dark places. At first there is not much effect on the appetite, but later the dog is inclined to eat indigestible objects, such as pieces of wood, and to eat its own feces and lap up its urine. The salivary secretion is increased and saliva hangs from the mouth in long strands. The dog will often bite or scratch the place of infection and may sometimes extensively injure the area. This stage lasts from 1 to 3 days.

2. *Maniacal or irritative stage.* In the stage of irritation the symptoms reach a higher degree of development. Excitement increases. Hallucinations which bring about violent fury or rage are apparent, and the voice of the animal changes decidedly from the normal. This is a very characteristic symptom of rabies and no doubt develops from the congestion and paralysis of the vocal cords. At this stage animals become vicious and, if confined, show a tendency to chew objects and tear them to pieces. If they are not confined, they will wander away and bite other animals or persons. Often, during a short period of time they will travel long distances, but they invariably return. A rabid animal at this stage, unless confined, may inoculate a large number of animals. As this stage progresses, the period of fury becomes shorter and marked depression is soon noted. The stage of fury lasts 2 to 4 days.

3. *Paralytic stage.* In the paralytic stage, the symptoms of paralysis become pronounced and the dog is less vicious. Complete loss of voice and inability to swallow, with paralysis of the tongue and chewing muscles, are early symptoms of this stage. The mouth is held open. The dried tongue is protruded and completely paralyzed. The eyes are

retracted, and lose their expression and luster; the pupils become dilated and in some cases unequal. The animal shows marked general wasting and much exhaustion. The general paralysis asserts itself, beginning in most cases in the tail and hindquarters. The animal has difficulty in walking or standing. Paralysis rapidly ascends to the foreparts and central nervous system, causing death on the fifth to eighth day of the attack. During the early stages of rabies the temperature may be high; later it becomes subnormal.

(*b*) *Dumb rabies.* This form of the disease is first characterized by paralysis of the lower jaw, tongue, and throat. This symptom develops early in the disease and is the most prominent one until general paralysis manifests itself. Often it may give the mistaken impression that the animal has a bone or some other object lodged between its teeth or in its throat. In the majority of cases, no nervous or excitable symptoms are shown; the paralytic symptoms gradually become more pronounced until the animal succumbs to general paralysis. There is the same peculiar expression in the eyes as in other rabid forms. The dog often attempts to eat and drink, but is prevented by the paralysis. Usually there is no tendency to biting or other viciousness. On the contrary, the dog tends to become more affectionate. The disease is fatal in practically all cases.

(3) *Treatment.* After rabies has developed in an animal, no treatment will have any affect on the disease. Under no circumstances should a dog showing symptoms of rabies be killed. It should be securely confined in quarantine until the disease has terminated fatally. Immediately thereafter the head should be cautiously severed, placed in a watertight container, packed in ice (the ice should be replenished as necessary), and sent to a competent laboratory for microscopic examination. Wounds caused by the bite of an animal known to be or suspected of being rabid should be opened freely, washed

thoroughly with an antiseptic, and cauterized with fuming nitric acid. Bitten animals will be held in quarantine for such period as is recommended by the veterinary officer.

(4) *Prophylaxis.* All dogs should be vaccinated with rabies vaccine. This vaccination is repeated annually and at such subsequent periods as are recommended by the veterinarian.

e. Mange. Mange is a communicable parasitic skin disease spread by both direct and indirect contact.

(1) *Cause.* Mange is caused by a very small animal parasite which lives on or in the skin.

(2) *Symptoms.* There is a patchy loss of hair associated with intense itching. The surfaces of the affected areas are likely to be moist and raw or scabby because of the activity of the parasite and the scratching or rubbing by the dog. Later the skin becomes much thickened and tends to lie in ridges.

(3) *Treatment.* Mange is not easily cured. It requires thorough treatment. Infected dogs must be isolated and their kennels thoroughly disinfected. The skin of the animal should be cleaned and all crusts and scales removed. This may be done by washing or scrubbing with water containing a mild soap. One of the best preparations for use in the treatment of mange is a lime-sulphur dip containing 1½ percent of polysulphide sulphur. As a rule, at least three treatments are necessary at 48-hour intervals; and in very advanced cases six or more may have to be given. The solution should be heated to at least body temperature and the animal allowed to remain in the bath for 5 minutes, in order to attain the maximum results. Precautions should be taken to prevent the solution from running into the ears or coming into contact with the eyes.

f. Ringworm. This is a communicable skin disease spread by direct or indirect contact.

(1) *Cause.* Ringworm is caused by a fungus similar to ordinary mold. This fungus will remain infectious in kennels and on equipment for a long period.

(2) *Symptoms.* Ringworm is characterized by small,

well-defined circular areas, occurring most commonly on the head and legs. These areas are first slightly reddened, but are soon covered with dirty, gray, crusted scabs. At times there is intense itching, inducing scratching and rubbing which remove the crusts and leave surface raw and bleeding.

(3) *Treatment.* In cases where large areas are involved, dipping in hot lime and sulphur is recommended. Small areas may be treated with tincture of iodine. Patients must be isolated in quarantine, and all personnel handling these cases must exercise precautions to prevent contracting the disease.

40. NONCOMMUNICABLE DISEASES. a. Eczema. Eczema is a noncommunicable disease of the skin that resembles mange in appearance. It may be caused by improper grooming, fleas, lice, irritation from the harness, worms, lack of exercise, or faulty diet. It is quite common among sledge dogs and requires patient and thorough treatment.

(1) *Symptoms.* Early symptoms reveal redness of the skin at the base of the tail, on the flanks and along the abdomen, accompanied by intense itching. As the condition progresses, there is a patchy loss of hair. The affected areas may appear moist, raw, or scabby from intense scratching. Later the skin is bare and thick and tends to lie in folds.

(2) *Treatment.* The dog should be thoroughly groomed and all matted hair removed. If weather conditions permit, wash the dog with a mild soap and water. All crusts and scales should be removed. Dip the dog in a lime-sulphur solution containing 1½ percent polysulphide sulphur (1½ ounce of sulphur to a gallon of warm water). Heat this solution to body temperature and place the animal in the solution for 5 minutes; or apply the liquid to the affected parts with swabs of cotton or a sponge. Care should be taken to avoid the eyes. Treat mild cases twice weekly and severe cases every second or third day until the condition is relieved.

(3) *Prevention.* A proper diet, daily grooming, pre-

vention of fleas and lice, proper kennel sanitation, and maintenance of harness in good repair will aid in the control of eczema.

b. Frostbite. (1) Lightly furred dogs are subject to frost bite on the tips of the ears, pads of the feet, testicles, and underparts of the thighs.

(2) *Treatment.* Put the dog in a cool sheltered inclosure to thaw out. Place the hands on the affected parts to stimulate recovery, but *do not rub or massage the affected parts.* Eskimos sometimes wrap the frozen thighs of dogs in fur skins.

c. Frozen lungs. This condition sometimes results from working dogs hard in extremely cold weather.

(1) *Symptoms.* Difficult breathing, wheezing, coughing, collapse.

(2) *Treatment.* The dog should be placed in a sheltered inclosure and wrapped in blankets or a protective covering. If the dog is stricken while on the trail, the animal should be placed on the sledge and transported to the nearest station. No attempt should be made to work the dog until it appears to be normal. Dogs that do not show signs of recovery in a few hours should be treated in the same manner as dogs with pneumonia.

41. DISTURBANCES OF THE DIGESTIVE TRACT. a. Diarrhea. (1) This term is applied to a too frequent evacuation of feces of a more or less liquid nature. It may be caused by drinking excessive quantities of milk or water, by eating large quantities of fat meat or salty fish, or by over-exertion.

(2) *Treatment.* Withhold all food from 24 to 48 hours. Administer two to four tablespoonfuls of castor oil. If diarrhea persists after this treatment, limit the amount of the animal's drinking water and give bismuth subnitrate in 10-grain doses every 3 hours until the condition subsides. Do not administer worm medicine.

b. Persistent vomiting and diarrhea. When these conditions exist, withhold all food and water for 12 hours. If the dog is extremely thirsty, it may be permitted to eat small quantities of snow. Neither worm medicine nor purgatives should be administered until the cause of the condition can be determined. Administer bismuth as in the treatment of diarrhea. When the dog has recovered, allow small quantities of food at 3-hour intervals. Two tablespoonfuls of mineral oil morning and night for 2 or 3 days will soothe the bowel and act as a laxative.

c. Constipation. (1) This condition usually results from heavy feeding, insufficient exercise, and the use of bones and dry biscuit in the diet.

(2) *Treatment.* To encourage elimination, make a suppository from a piece of noncaustic soap, and insert it into the rectum. Allow the dog to drink plenty of water, and eliminate bones and biscuit from the diet. Two tablespoonfuls of mineral oil, morning and night for 2 or 3 days, will prove beneficial.

d. Poisoning. (1) When it is known or suspected that a dog has been poisoned, vomiting should be induced immediately.

(2) *Treatment.* Feed the dog warm soap water or warm salt water in large quantities to encourage vomiting. When the dog has vomited freely, give it a heaping tablespoonful of epsom salts in a small quantity of water. Keep the dog warm and administer large quantities of milk. If possible, seek further medical treatment at the nearest station. When the animal appears to have recovered, it should be rested for several days and fed small quantities of milk and lean meat.

e. Foreign body in the mouth. (1) When dogs claw at the mouth, have difficulty in eating, or have an excess flow of saliva, the presence of foreign objects in the mouth should be suspected. Pieces of bone, splinters of wood and small stones often lodge between the teeth and cause distress.

(2) *Treatment.* Removal of the foreign body gives immediate relief. Caution must be exercised when a dog exhibits symptoms of this nature, as such symptoms are noted also in dogs having rabies. If the dog has previously been bitten by a stray animal, do not attempt to examine the mouth. Isolate the dog and endeavor to

diagnose symptoms. Consult competent authorities if available.

42. DISTURBANCES OF THE NERVOUS SYSTEM—FITS. The causes of convulsions, fits, or hysteria may vary from digestive disturbances to organisms causing infectious diseases. When a dog has a fit, do not attempt to apply restraint or administer medicine. If the dog is on a leash, tie him to a stake or place him in a sheltered inclosure until the condition subsides. When the condition has passed, try to determine the cause and then apply the indicated treatment. When this condition occurs in mature dogs, it is advisable to withhold food and administer a purgative.

43. INTESTINAL PARASITES. Dogs are not infrequently infested with worms. The most common types are tapeworm, roundworm, and hookworm.

a. Tapeworm. (1) This type of worm is common to dogs of all ages, and the symptoms it produces vary considerably. The flea is required as an intermediate host for completion of the life cycle of the tapeworm. Affected dogs often show symptoms of decreased appetite, or may become greedy, develop a shaggy coat, and grow thin. A correct diagnosis of tapeworm can be made only by finding worm segments, which resemble grains of rice, attached to the hair around the rectum or on the dog's bedding. Tapeworms are acquired by eating raw or insufficiently cooked flesh of animals and fish, and occasionally from swallowing infested fleas.

(2) *Treatment.* Administer arecoline hydrobromine in doses of 1/10th to 1 grain, depending on the size of the dog. Usually ¼ grain is sufficient for a 40-pound dog. The drug is administered preferably in a capsule and is well tolerated, but may easily cause vomiting. In order to prevent vomiting, dogs should be prepared for worming by drinking only milk for 24 hours and then fasting for 24 hours. Usually the worms are passed 10 to 20 minutes after administering the medicine. Severe

cases of tapeworm infestation may require further treatment at monthly intervals.

b. Roundworms and hookworms. Both types of worms are common in young dogs, especially in pups. The source of infestation is the eggs of female worms passed in the feces of an infected dog. Under natural conditions, the development of the embryo may take place in water or soil. The eggs are extremely resistant and may remain viable for a considerable period. Dogs become infested by ingesting these embryonic eggs with food or water. Infestation with these parasites commonly occurs in unborn pups, which accounts for the presence of worm infestation in newborn pups.

(1) *Symptoms.* The pups show symptoms of wasting at from 3 to 5 weeks of age. They are stupid and do not play as animals usually do at this stage. Vomiting is common, and quite often some of the worms are expelled. Leanness increases, the linings of the mouth and lids are pale, and the abdomen is enlarged. Diarrhea is not uncommon, often alternating with constipation. Frequently the symptoms become more severe and food is refused; this is followed by weakness, unconsciousness, and death. In mild cases, or in older dogs, the symptoms are less severe and often remain unnoticed. The appetite remains good and is often voracious. The hair coat is dull and rough, and its growth is impaired. Positive diagnosis can be made only by finding the eggs in the feces, or by finding worms in the vomitus or feces.

(2) *Treatment.* Treatment should be given as soon as the symptoms are noted and before the dog becomes too weak. Oil of chenopodium, in doses of 1 to four drops in ½ to 1 ounce of castor oil, produces excellent results in pups. This should be given after a short period of starvation. Mature dogs are treated with tetrachlorethyle, which is furnished in 1-cc capsules. Fast dogs for 18 hours, and administer three 1-cc capsules for a 50-pound dog. Do not feed fat or fatty foods for 48 hours before or after administering the drug. A laxative dose of epsom salts (one tablespoonful dissolved in water) should be given 2 hours after the capsules are administered. Follow 2

hours later with small quantities of food. Repeat treatment after one month, if animals continue passing worms in the stool.

(3) *Prevention.* In order to rear young animals free from roundworm and hookworm infestation, energetic measures must be directed toward a cure of all infested dogs and a thorough disinfection of all premises to which they have access. This is best done in the early winter, because the weather conditions then are unfavorable to the development of the parasites. All new arrivals to the kennel should be examined and treated before being allowed with the other dogs. Pregnant animals should be entirely free of all parasites and thoroughly washed to remove all eggs or embryos that may be on their hair or skin. Feed and water pans and all receptacles must be protected against soiling by dogs that harbor worms.

44. EXTERNAL PARASITES. Fleas, lice, and ticks are parasites found on the body of the dog. In cold climates, sledge dogs are not usually troubled with fleas and ticks. Lice, however, may be a problem.

a. Fleas. The flea is a jumping insect without wings, which attaches itself to the dog and obtains nourishment by sucking blood. It is brown in color, can easily be seen, and causes the dog considerable annoyance. Flea eggs are deposited on the floor or ground and hatched in 6 to 12 days. They develop into adult insects in 3 to 5 weeks, depending upon the temperature.

(1) *Symptoms.* Feeble or young animals are particularly susceptible, as are animals confined in close quarters. Fleas harm the animal chiefly by disturbing its rest. Large numbers may produce chronic skin eruptions, but the greatest damage to the skin results from constant scratching, which produces irritated areas on the neck, back of the ears, and back of the forearms.

(2) *Treatment and control.* As fleas live in unsanitary surroundings, thorough grooming and strict sanitation and disinfection are required for their elimination. The destruction of fleas on the dog may be effected by the

application of flea powder or insect sprays, or by bathing the dog in a 2 percent solution of creolin.

b. Lice. Two varieties of lice infest dogs: biting and sucking lice. The dog louse is white in color, but will assume a red color when filled with blood. The "nit", or eggs of the louse, will be found firmly attached to the dog's hair.

(1) *Symptoms.* Lice do not seem to annoy dogs as much as do fleas. A mild irritation of the skin results from scratching the irritated parts and from bites inflicted by lice.

(2) *Control and treatment.* Cleanliness discourages lice propagation. When dogs are found to have lice, they should be isolated until free of the parasite. All dogs in the kennel should be groomed daily to detect further evidence of the condition. Kennels should be thoroughly washed and disinfected and all bedding destroyed by burning. Dogs infested with lice should be clipped if weather conditions permit, and the hair should be destroyed by burning. The animal should then be bathed in a 2 percent solution of creolin. If clipping is not practicable, the dog should be groomed daily and sprayed with an insecticide. Affected dogs should be isolated and all loose hair should be burned.

45. WOUNDS AND INJURIES. a. Wounds. A wound is any break in the skin, the body tissues, or the lining of a body cavity, resulting from external violence or muscular activity of the body itself. Wounds are classified as incised wounds or cuts, lacerated wounds or tears, and puncture wounds or holes. The types of wounds most frequently encountered in sledge dogs are tear and puncture wounds resulting from fighting.

(1) *Treatment.* (*a*) If bleeding is present, it must be controlled before any attempt is made to clean or bandage the wound.

(*b*) Bleeding is controlled by the application of pressure or of a tourniquet directly to the bleeding surface. A tourniquet may be applied to the legs or tail to control a hemorrhage. It may be fashioned from a strip of gauze bandage, a handerchief or a necktie, and should be placed on the side of the wound nearer the body. Loosen the tourniquet at 15-minute intervals to relieve pressure, but do not remove it. When bleeding has been controlled, clean and dress the wound. Should the bleeding continue, tighten the tourniquet again. When it is not possible to apply a tourniquet, bleeding may be controlled by pressure applied directly to the bleeding surface. Do this by covering the bleeding surface with a clean piece of gauze or a handkerchief to make a compress over the wound, binding it firmly in position with a bandage.

(*c*) When bleeding has been controlled, clean the wound. The hair should be clipped with scissors and the dirt washed away from the injured parts with clean water. All visible foreign material should be removed from the wound, but the wound should nol be packed, probed, or covered with ointment. If available, an antiseptic may be applied and the wound covered first with gauze, then with cotton, and then bandaged. Never place cotton directly over a wound, and never apply a bandage over a tourniquet.

(2) *Closing of wounds.* Large gaping wounds should be closed with stitches and bandaged. The bandage should be firmly fastened with adhesive tape to prevent the dog from tearing out the stitches. In applying stitches, the edges of the wound should be brought together in their natural position. The needle and thread are passed through the skin at one side of the wound and out at the other, and tied so as to approximate the edges without tension. Stitches should be from 1/4 to 1/2 inch from the edge, and about 3/4 inch apart. The stitches may be removed in 8 or 10 days.

(3) *Drainage.* Should infection develop, several of the lower stitches should be removed and the edges of the wound spread to allow for drainage.

(4) *Dressings.* After a wound has begun to fill in with repair tissue, it will often heal much better if no dressing or bandage is placed in contact with it, provided the dog does not attempt to chew the wound and it is not irritated by insects. It is preferable to rest a dog until

wounds are healed. All wounds should be kept dry and dressings should be changed only often enough to keep the wound clean. Little washing as possible should be done and the wounded parts should be sopped instead of rubbed.

b. Injuries. Injuries are classified as bruises, strains, sprains, and fractures. They are caused by external violence or by muscular activity of the body itself, and involve muscles, tendons, ligaments, joints, organs, and bones. A break in the tissue covering the body need not accompany an injury.

(1) *Bruises.* Bruises are injuries without a break in the overlying skin. They are frequently caused by a blow from a blunt object.

(*a*) *Symptoms.* When bruises occur over a muscle, they produce soreness and lameness. If severe, they may rupture a blood vessel which will result in a fluctuating swelling caused by accumulation of blood, usually just under the skin.

(*b*) *Treatment.* Rest and hot applications are required. In bruises where there is a fluctuating swelling that does not reduce in 7 to 10 days, the swelling may be opened at its lowest point with a knife, to allow the fluid contents to escape. After opening, the incision should be kept clean and painted with iodine once daily.

(2) *Strains.* A strain is an injury to one of the larger muscles or muscle groups. It results from stretching or tearing a muscle by sudden and violent muscular action.

(*a*) *Symptoms.* Pain, tenderness, swelling, and disability occur in the region affected.

(*b*) *Treatment.* Hot applications and rest are required.

(3) *Sprains.* A sprain is an injury to a joint which does not involve a break in the overlying skin. It is caused by twisting or pulling the joint or by bending the joint beyond its normal range of action, which results in stretching or tearing of the ligaments.

(*a*) *Symptoms.* Marked lameness, heat and swelling over the joint, and tenderness on pressure or movement are the signs.

(*b*) *Treatment.* Rest, hot applications, and massage with chloroform liniment once or twice a day are indicated.

(4) *Fractures.* A fracture is a break in a bone. Fractures are classified as *simple* and *compound.* In a simple fracture there is no break in the skin, while in compound fractures a fragment of bone protrudes through the skin. Normally a broken bone does not necessitate destruction of the dog. However, if the injury is received on the trail and it is impossible to carry or rest the injured animal, destruction may be necessary.

(*a*) *Symptoms.* When one of the long bones of the leg is completely fractured, the dog evidences considerable pain and the leg dangles helplessly. In incomplete fractures there is usually swelling and tenderness at the site of the break and the limb will bear no weight. Crepitation (grating of the broken ends of the bone) may be felt by placing the fingers above and below the injured part and gently moving the leg.

(*b*) *Treatment.* Before attempting to administer first aid, put a muzzle on the dog. Then, assuming that the dog has sustained a leg injury, place the dog in a reclining position with the injured limb away from the ground, and carefully examine the leg for the extent of the injury. When the leg is found to be fractured (broken), or a fracture is suspected, obtain two flat pieces of wood or heavy cardboard, and fashion splints the length and shape of the leg. If help is available, have an assistant place one hand under the injured part and grasp the toes with his other hand. Allow the leg to remain in its normal position, cover it completely with a layer of cotton, place one splint on the inside and the other on the outside of the leg, and securely bind the entire length of the splint to the leg with rolls of bandage, maintaining an even pressure along the entire length of the splint. Adhesive tape is then firmly applied over the bandage. Injuries are accompanied by swelling, and tight bandages will cause the dog unnecessary distress. Pad the foot with cotton, but do not include it in the splint unless the foot is fractured. A leg that has been bandaged too tightly will evidence swelling of the foot and at the top of the splint. When swelling occurs, the bandage must be released. In compound fractures, clip the hair, clean the wound, apply an antiseptic, and gently move the broken bone back under the skin. With the leg in normal position, apply splints, bandage, and tape, leaving a window (opening) in the cast over the wound so that it can be cleaned and dressed daily. After dressing the wound, the cellophane wrapper from a cigarette package may be fixed with tape over the opening in the cast to prevent the entrance of dirt. In a compound fracture, the splint should remain in position for 6 to 8 weeks, while with simple fractures splints may be removed after 3 weeks.

(5) *Eye injuries.* Injuries to the eye are very common, and result usually from violent contact with foreign objects.

(*a*) *Treatment.* In treating an eye injury, the eye should be bathed with clear warm water and then examined for the presence of any foreign material. Foreign material should be carefully removed, provided removal will not cause further injury. A moist piece of gauze should then be placed next to the eye, and cotton and bandage applied.

(*b*) Occasionally a dog's eye will be prolapsed or forced outside the lids. Dogs with prominent eyes are predisposed to this condition which may result from injury or fighting. An attempt should be made to replace the eye immediately. This is done by grasping the lower lid firmly with the thumb and fingers and pulling outward. At the same time place a piece of clean gauze, moistened with mineral oil, on the eyeball and push steadily and firmly inward. Usually the eyeball will slip back into position. If this cannot be achieved within a few minutes, moisten the eyeball thoroughly with mineral oil, cover with gauze and cotton, and apply a bandage.

(6) *Bandaging.* (*a*) *Uses.* Bandages are employed under varying conditions for many purposes. The following are their chief uses:

1. To control bleeding by pressure.

2. To hold dressings and compresses in place.

3. To hold splints in place.

4. To immobilize a part of the body.

(*b*) *Application.* The roller bandage is the type most frequently used for applying dressings, splints, compresses, and pressure.

1. In applying a bandage, use one hand to hold and direct the roll and the other to keep each lap smooth and to hold the turn while taking up slack or changing direction. Apply the bandage to the injured part snugly enough to exert the desired amount of pressure and to prevent slipping, but be careful not to shut off circulation by too much pressure. When the ends of the bandage are secured they should be tied no more tightly than the bandage itself has been bound.
2. A bandage should be applied over gauze and cotton and not directly to the body. It is not advisable to apply a wet gauze bandage, because in drying it may tighten and cause undue pressure. Bruised areas usually become swollen within a short time following injury. This should be remembered when bandaging, as the application of a snug bandage to the original wound will impair circulation if swelling occurs.
3. In applying a bandage to the leg, it is advisable not to bandage the foot unless it is involved in the injury. Tight bandaging will cause swelling of the foot, and this condition can be quickly recognized and corrected if the foot is exposed. In securing the ends of a bandage, fasten them on the outside of the leg, but not directly over the injured area; pressure from a knot on the wound will cause the animal discomfort.
4. When it is necessary to apply a bandage to the body itself, a many-tailed bandage is effective. This is made from a piece of cloth long

enough to encircle the body completely and to permit the making of ties as well. The material is folded in half and strips are cut into it at the open ends. The dressing is placed on the chest, flank, or abdomen, the bandage placed around the body, and the strips tied on top of the back. Made on a smaller scale, a many-tailed bandage may be applied to the head or neck.

(7) *Restraint of dogs.* (*a*) Before administering first aid to the injured dog, it is always advisable to place a muzzle around the dog's mouth, no matter how well the soldier is acquainted with the normal behavior of the dog. It must be remembered that an injured or frightened dog will often bite instinctively in self-defense.

(*b*) The muzzle is fashioned from a piece of gauze bandage or from a firm strip of cloth. It is placed around the mouth and looped under the lower jaw. The ends of of the bandage are pulled taut to tighten the loop, and then passed to the back of the neck just below the ears, where the ends are securely tied.

(8) *Dew claws.* Dew claws on the inside of the hind legs should be removed. They serve no useful purpose and are subject to injury when dogs break through crusted snow. Dew claws may be removed easily with a pair of sterilized, curved scissors, when pups are 4 or 5 days old. When present on older dogs, they should be removed surgically by a veterinarian.

(9) *Toenails.* Ordinarily the nails wear down naturally, but if the dogs are used in soft snow for long periods the nails may require cutting. Inspection should be made regularly and the hard ends of the nails cut off when necessary. Long nails often break or grow into the pads of the feet and disable the dog.

(10) *First-aid supplies.* For use in treating dogs in the absence of veterinary service, or for first-aid treatment, small quantities of medicine, bandages and adhesive tape may be issued. Requisitions for the necessary medical supplies should be submitted through proper military channels.

CHAPTER 4

INDIVIDUAL AND TEAM TRAINING

---

46. GENERAL. Dog sledgers generally have their own individual ideas on how to meet varying conditions of weather and terrain, and this results in wide divergence in training methods and sledging techniques. Customs handed down by various Eskimo tribes also have considerable influence on training methods. Wherever possible, use should be made of the knowledge and experience of Eskimo dog drivers. In spite of individual theories and practices, certain fundamental principles of dog training have been tried and proved. These are discussed in this chapter.

47. COMMANDS. a. A great variety of commands may be used to drive a team, and the dogs will respond to any words they hear regularly. Starting commands such as "Yake" have been used in the Antarctic; "Mush" is a corruption of the French word "Marche", and is used in parts of Alaska and Canada; and the simple words "Hike" and "All right" have also been found effective as starting commands. The team should be trained to sit down or even to lie down in the traces as the command "Down". Good lead dogs learn quite readily to respond to a few simple commands.

b. The following are Army commands which are almost universally used:

(1) "Yake" or "All right", for starting.
(2) "Gee", for a right turn.
(3) "Haw", for a left turn.
(4) "Whoa", for stopping.
(5) "Haw come in", for a complete turn around.

c. In Greenland, dogs and drivers are accustomed to

different commands. The order for turning right is "Ille", spoken in a high-pitched voice. "Yuk", spoken in a nasal voice, is the command for a left turn. To stop the team, the driver whistles, or says "Aye".

48. INDIVIDUAL TRAINING. a. Characteristics and duties of the lead dog. (1) A good leader is of prime importance in a well-trained team, and much care and attention should be given to the training of this dog. The time spent will be well rewarded by the tremendous amount of work and trouble a good leader saves a driver. Many drivers claim that a bitch makes the best leader, but this does not always hold true. Leadership is a natural gift in most instances, and a dog must have certain definite qualities to be a leader. Obedience is one of the chief requisites, for without it commands are useless. In addition, the leader must possess speed, intelligence, and a considerable amount of natural authority. Close observation and continued handling of groups of dogs for several weeks will reveal the animals that possess these distinctive qualities.

(2) The importance of the leader is evident from the fact that there is no rein on the team. Dogs will follow a well-broken trail over a familiar course, but when it is necessary to turn, or when traveling over open country, a lead dog responsive to commands is a necessity.

(3) More than one leader should be available in each team. The best leader will be used regularly and the others driven in point positions in the team until emergency requires their use. Additional leaders can be used to take their turn breaking trail.

(4) In addition to obeying commands, a leader is expected to "hold out" the towline. The dogs are attached to this line by their neck chains in front and traces at the rear, and of necessity keep their respective places when the line is taut. It is an important responsibility of the leader to keep the team strung out in formation by "holding out" the line at all times on the trail and also while the team is being hitched and unhitched. A good pair of dogs immediately in his rear is desirable

to help him "hold out" the line. If the lead dog lies down during halts and when hitching and unhitching, the other dogs will generally follow his example.

(5) It is the responsibility of the lead dog to pilot the team past distractions. Upon meeting a team, he should turn out widely to lead his team around and thus avoid a melee. The team, fighting as a pack, will quickly dispatch a single dog that comes within reach. In open country, a good leader may be expected to lead a runaway team on a circular course back to the driver.

(6) Over open ground, and during storms when the trail is indistinct to the driver or has been obliterated, the lead dog is expected to follow it. He may do this by scent, running slightly on the leeward side of the old trail and zigzagging when he has lost it. When there is no trail, he is responsible for keeping the general direction indicated by the driver. The tendency to run straight into, or away from, a driving storm must be resisted when the trail takes an oblique course. Free (unharnessed) leaders may be employed. A good leader not in harness will keep his position at the head of the team and will have greater freedom of movement and be able to devote more strength to the task of breaking trail when he is relieved of the responsibility of holding-out the line and pulling a share of the load.

(7) Although a leader may be used in point position, it will discourage the good point dog or leader to use him in other than the point or lead positions.

b. Training the lead dog. (1) Various methods can be used effectively in training a lead dog. Walking with the prospective candidate on a leash may be one of the preliminary steps. The first thing to teach the dog is to march in front of the trainer with the leash taut. Then guide the dog with the leash as the commands are given. Taking hold of the dog's collar and guiding it as the commands are given is sometimes effective, but it is not recommended. Training on the leash should continue during snow-free periods. It is advisable to take the lead dog out as often as possible for such training during idle periods. This training can be carried out on foot or on skis. (See (3) below.)

(2) When a dog shows a natural willingness or eagerness to go forward and to assume a position in front, he has the makings of a lead animal. Such a dog should be placed as one of the two point dogs, directly in back of a trained leader. In this position he will receive a partial training as a lead dog. Sometimes it is advisable to hitch the trainee alone behind the leader; an even better practice is to hitch him with an experienced lead dog in a sort of double lead arrangement. When the latter method is used, hitch the trainee to a shorter tug rope, to place him in a position slightly in back of the experienced lead dog. At still other times, it may be advisable to release the experienced lead dog entirely from the towline. This dog will still keep out in front and will permit the trainee to act as the actual head of the team.

Figure 20. Ski-jori

(3) *Ski-joring* is also a very effective method of training a leader (fig. 20). The trainee is hitched with only one pair of dogs behind. The driver operates on skis and is merely towed behind the dogs, without a sledge. Only

experienced skiers should use this method. The towline can be held by means of a crossbar or a loop in the rope. The skis will provide for good control in case the commands are not properly executed.

(4) When first training the lead dog to respond to commands, it is imperative that each command taught be executed properly. Never permit the dog to take any direction or to perform any maneuver other than that called for in the command. Good control of the sledge (brakes, in the case of winter training with a sledge) is required to accomplish this training.

c. Training individual dogs. The two commands which must be obeyed by every dog in the team are "down" and "whoa"; the latter command is the more important.

(1) *"Down".* Each dog should be trained individually to obey the command "down". Training consists of tapping the animal's rump with your hand while the order is being given.

(2) *"Whoa".* Training for the command "whoa" should be given to each unit of two dogs, with the leader in front of them. Immediately after the command "whoa" has been given, the order "down" should be given and obeyed.

(3) Light pressure of the leash can be applied in the early stages of training. At a later stage, all orders should be obeyed with the leash slack.

d. Punishment. For full effectiveness, punishment should always be administered immediately a dog misbehaves. Punishment with the long whip while the sledge is in motion should be administered only if a driver is skillful enough to hit the dog he is aiming for. Otherwise, the offender should be unhooked and taken out of the team. A well-known and effective method is to grip his head firmly between your knees and, holding his muzzle in your left hand, give him several sharp blows on the nose. Then return him to his place. He should not be hit on the body.

49. UNIT TRAINING. To obtain best results, dogs

should be taught to work together as a unit. Except for occasional replacements, these dogs should always be kept together and should be trained as a team. Once the dogs become accustomed to each other, they will work better together and show less tendency to fight among themselves.

a. In forming the team, place the faster and more energetic dogs toward the front to help the leader hold the gangline taut. In the double tandem hitch (par. 57, fig. 21), the point pair of dogs (immediately behind the leader) should be fast and themselves potential leaders. The wheel dogs (immediately in front of the sledge) must be strong, powerful dogs capable of giving the sledge its initial start. Older and more experienced dogs should be used as wheel dogs, for pups will often become frightened when placed close to the sledge.

b. In selecting the dogs to make up a team, try to choose animals more or less evenly matched in size and weight. It is also desirable, though more difficult, to select dogs who have similar gaits. (Some dogs pace or trot, and others seem to have a lope.)

c. The same driver should handle the team all through the training period.

d. The long whip (par. 16) should be used in training an entire team only if the driver is adept in its use.

e. The team should proceed at a fast trot, whenever possible. The dogs should be dissuaded from galloping, however, for such a pace wears them out. Consideration of all these factors will contribute toward the formation of a smooth-working, efficient team.

50. HITCHING. **a.** During the hitching period the dogs are always eager to be off; it is therefore wise to secure the sledge before the dogs are brought out for hitching. This will eliminate runaways, which would otherwise be quite common. The method of securing the sledge is to attach a rope (the snubline) to the connection of the last section of the towline and the sledge bridle, and then to a convenient pole, with a slipknot. When the driver is ready to go, he simply slips the slipknot and is off. The sledge anchor (par. 13, fig. 14) may also be used for the same purpose.

b. An excited dog can drag a man by the chain, and footing on snow and ice is seldom secure. If a dog becomes excited and it is necessary to restrain him by lifting his feet off the ground, lift him by the neck strap.

c. Before hitching, the towline assembly, with harnesses attached, should be attached to the sledge and stretched out on the ground (fig. 10).

d. The lead dog should always be hitched first, and he must be trained to sit or lie quietly with the towline stretched taut so that the other dogs will be unable to mill about. After the lead dog is in place, the quieter dogs should be hitched first and the more pugnacious ones last. It is good training to make all the dogs lie down before the start.

51. UNHITCHING. **a.** Before unhitching, all the dogs should be made to lie down and then the sledge should be secured.

b. The lead dog should be the last one unhitched, and the others should be taken in inverse order of hitching.

c. The dogs should be examined for sore feet and treated for any other injuries.

d. The sledge should then be unloaded, and, after scraping and wiping the runners, leaned against a tree. In barren country, the sledge should be turned on its back to keep ice from forming on the runners.

52. TRAINING THE PUP. **a.** The training of sledge dogs should begin at 6 or 7 months of age. The usual procedure is first to let the pup run loose with the team on a few trips. The next step is to harness a small, slow, steady team (five to seven dogs), including if possible the mother of the pup which is to be trained. One pup should be trained at a time. It should be harnessed and put in the team, next to the mother, and toward the rear or in the middle of the team. Never put a pup at wheel; its actions are unpredictable. It may lie down, run under the sledge, or pull eagerly along with the rest. Patience and self-control on the part of the trainer at this stage are essential. Never let the team outrun the pup so that it cannot keep up. If the pup should lose its feet and be dragged forward by the others, stop the team immediately. Many pups have been ruined by fright arising out of their first sledging experiences. Should the pup insist on dragging in the harness, it may be induced to get up and run by a very light lash or tap of the hand on the hindquarters. If this fails, let the pup run alongside the team until it catches the pulling spirit. An extra man riding on the sledge provides a light load, which is desirable; he can help also to untangle the pups and keep the team in order.

b. Most sledge dog pups are eager to pull once they get the idea. After one pup starts to pull, hitch another pup alongside it; but never train more than three pups together in a seven or nine dog team, until they have become accustomed to pulling. Care should be taken not to give the pups heavy work. At the beginning they should not pull more than half an hour at a time. A pup should not be expected to do a full day's work until it is at least 16 months old.

53. USING THE WHIP. **a.** A whip should be carried on the sledge at all times, more as an insurance against fighting than for actual whipping. At times it may be necessary to use the whip, for instance, when two teams pass each other, when fights start within a team, or when a dog goes lazy in harness. The whip must be used judiciously at all times and in such a manner that the dogs will suffer no injury. Unfortunately, many drivers have a tendency to become so angry in punishing the dogs that they lose good judgment. Constant use of the whip has no place in dog sledging, and is more often than not an indication of the driver's ineptitude in handling his team.

b. A long whip (of the West Greenland type) may be used to get maximum efficiency from a team in the fan hitch, provided the driver knows how to handle it with precision. (When the double or single tandem hitch is used, short whips are best.) The whip should be from 15 to 20 feet in length (par. 16). With a whip of this length, a lazy or disobedient dog can be called to attention without stopping the sledge. Furthermore, immediate punishment can be given with the long whip, and the dogs will thus learn at once that they cannot be lazy or disobedient.

c. Calling an individual dog by name, or brandishing the whip, is often sufficient to prevent fighting. Once a fight starts, stop it by some means as quickly as possible. A light sting on the end of the nose is very effective, but be careful not to injure the animal's eyes. Never kick a dog, and never give it a blow on the back of the ribs. A blow with the whip across the rump or shoulders may prove effective, but a heavy coat of fur gives good protection from the lash, and fighting may not be stopped by such a blow. It is dangerous to mix into a general fight without exercising great care. However, by catching hold of the dog's neck strap from the rear, raising his front feet, and giving him several raps on the nose, a driver can make a single dog stop fighting without incurring much danger.

54. PETTING. **a.** Affectionate dogs looking for caresses will not keep their positions, will disorganize the work of the team, and will take advantage of their driver. Pets will not maintain the businesslike attitude necessary on the part of both men and dogs when at work.

b. The average sledge dog will not display affection unless it is first invited by the driver. Good results will be attained when there is a feeling of mutual confidence and respect. A friendly attitude, as distinguished from the affectionate relationship with a pet, is desirable. The time for making friends is when the dog is not at work, but never before he is thoroughly trained in harness. This is especially true of the leader. It is difficult to keep him properly disciplined while at the same time making a pet of him.

c. Exceptions to the rule that sledge dogs should not be made into pets are common. Through long association, a close bond of feeling and understanding is created between some capable drivers and their teams. Examples are not uncommon of runaway teams and individual loose dogs that have returned to a driver for whom they have affection. However, a good rule in a kennel is not to pet the dogs, and to touch them only to hitch and unhitch them, or when necessary for other purposes. On the other hand, it is a mistake to assume that harshness and rough handling will get good results. Cruelty is out of place at all times.

55. FEMALES IN TEAM. Many drivers claim that females should be kept out of teams, but there appears to be no really valid basis for this attitude. Some polar expeditions have found that "all-girl" teams work very well together, especially for rapid travel when the loads are light and the distances to be covered relatively short. Females are often steadier workers than males, and the presence of a few bitches in a team will tend to quiet the more pugnacious dogs. A bitch in season will, however, disrupt the entire team; she should be kept penned up at this time.

CHAPTER 5

ON TRAIL

Section I. HITCHES

56. GENERAL. There is no uniformity among sledgers with regard to the use of hitches. In the United States, Alaska, and most of Canada, the *double tandem hitch* is mainly employed. All the United States Antarctic expeditions have used this hitch. The Eskimos in Greenland and the eastern Canadian arctic use the *fan hitch,* while the *single tandem hitch* is used in other parts of Canada and in Alaska, particularly in wooded regions. Each type of hitch has its own special merit; under different sledging conditions, one hitch may be more effective than another. It is advisable to train dogs to the use of as many of the different hitches as possible.

57. DOUBLE TANDEM HITCH (CENTER TRACE). a. The double tandem hitch (fig. 21), known also as the "Nome", or center trace, is the most versatile and in general will be the most widely used. The dogs are hitched in pairs on 68-inch sections of a center trace or towline. The advantages of this hitch are: all the power is exerted in one forward pull; only two trails are broken, and the sledge runners follow in these; there is less likelihood that the tug ropes will become entangled. Some drivers, however, find it more difficult to keep the dogs from idling with the double tandem than with the fan hitch.

b. The double tandem is the best all-around hitch to use in open country; it is good even where there is jagged ice, because the short tug ropes do not become snagged on sharp pieces of ice, as is often the case with the long

Figure 21. Double tandem hitch.

trace or tug lines. Opinion varies widely as to the type of hitch preferable for travel in crevassed regions, the choice being between double tandem and fan. The opinion of all the drivers on the Byrd and United States Antarctic expeditions is that there is a safety element involved in the use of the double tandem hitch for crossing bridges over crevasses. Once several of the front dogs get across these crevasses, they will help pull the others over. On a single long tug rope a dog may fall in by himself; such an accident might necessitate stopping the sledge in the middle of a crevasse to extricate the dog. The double tandem hitch permits less freedom than the fan type, but additional freedom may be obtained by unsnapping the neck lines.

58. SINGLE TANDEM HITCH. a. With the single tandem hitch (fig. 22), dogs are hitched in a single line or file. This arrangement has certain advantages in wooded country such as is encountered in parts of Alaska and northern Canada, where trails may be narrow and maneuvering must be done between trees. In such a case a narrow 17-inch sledge or toboggan is usually required with this type of hitch. Also in deep snow, where it is necessary to break only a single track, the single tandem has an advantage over other hitches. This hitch provides good results when only about five dogs are needed; otherwise it is not recommended.

b. Disadvantages. (1) Heavy loads on a single tandem hitch require a long file of dogs. This means a waste of pulling power.

(2) The long single line should never be used across crevasses.

(3) When the sledge is going over small ridges or knolls, the downward strain from the front dogs, after they have passed over the rise, presses the rear dogs down to the ground. This results in their carrying a heavy back load in addition to their forward pull. Similarly, in crossing a draw, the dogs in the middle of a long, single line tend to be lifted off their feet by the forward dogs when they reach the top of the draw.

Figure 22. Single tandem hitch showing use of "ouija board".

(4) The single tandem hitch requires a special harness.

59. FAN HITCH (TRUE FAN). a. The true fan hitch (fig. 23) has been used by many Arctic explorers for sledging in the North, and is used in treeless areas. Each dog has an individual tug rope, about 18 feet long from the harness to the sledge. These tugs are of equal length.

b. There should be one bitch in every team to insure a high spirit among the dogs. The king dog and the next strongest dog take up positions on either side of the bitch; the rest of the dogs will prefer one flank or the other.

c. The wings are key spots, and the wing dogs should therefore be chosen for their obedience to commands. The command to go should not be given until the important wing dogs are in position.

d. The tendency to change places can be curbed with the whip, as can any slackness on the part of the dogs.

e. The fan hitch should be used with dogs accustomed to it (Greenland Eskimo dogs in particular), and only in open country.

f. Advantages. The advantages of this hitch are:

(1) Each dog has more freedom than when hitched in tandem or double tandem and can pick his own way as well as jump more easily over crevasses or across leads.

(2) At least some of the dogs are able to get good footholds on rough terrain.

(3) It is possible to cut the dogs loose quickly in an emergency.

g. Disadvantages. The following are some disadvantages of the fan hitch:

(1) It is impractical for use in deep snow, as each dog would have to break his own trail, and would soon tire.

(2) The tug ropes can become badly mixed up in short order, particularly if the wing dogs start running inward.

(3) The flank dogs lose considerable power because of the angle at which they have to pull.

(4) The long tug ropes become snagged easily on sharp, jagged ice.

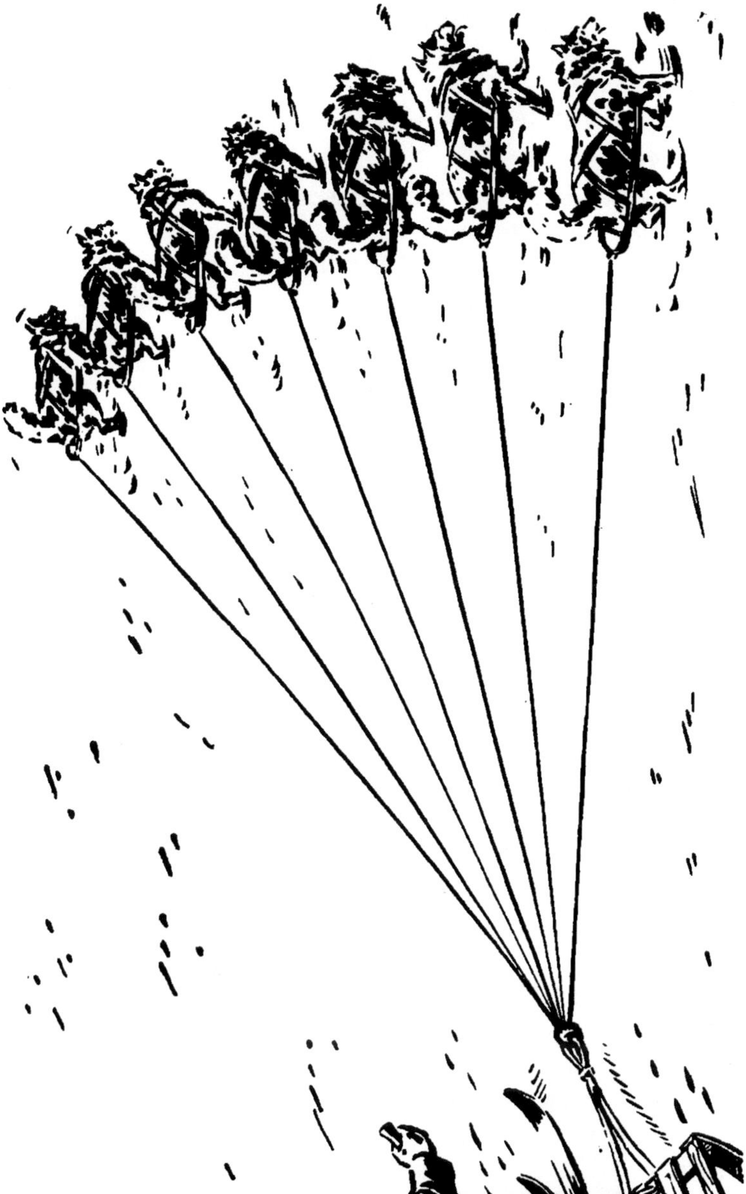

Figure 23. Fan hitch (true fan).

60. MODIFIED FAN HITCH. This system of hitching (fig. 24) is a modification of the true fan. Sometimes known as the Mackenzie River method, it has been used in the western and eastern Canadian Arctic, in the Antarctic by the British, and in Greenland and Lapland by the French.

a. The method consists of driving the dogs in pairs, but with each on a long separate tug rope to the sledge. A 6-foot leading tug may be attached to the front of the sledge and all the individual tugs secured to it. The center dog is farthest out in front, with the rest of the pairs spaced at intervals of 6 feet behind. The rear dogs are on 12-foot lead-lines, which means they are 18 feet from the sledge. This hitch which resembles the double tandem hitch, permits the dogs to close in more than is permitted by the true fan.

b. Advantages. The modified fan hitch overcomes some of the undesirable features of both the true fan and the double tandem.

(1) It gives the dogs companionship, which is fine for morale on long journeys.

(2) It enables the dogs to spread out on a good surface, or to assemble in double tandem style on a difficult one.

(3) It is an excellent hitch for use over leads, crevasses, between ice floes, and on hummocky ice (although on this latter surface the tug ropes may become snagged).

(4) It is also, in common with the other fan hitches, a safer hitch than the tandem or double tandem for travel on thin ice, since the dogs can spread out widely, and this means that there is less likelihood that all of them will go through the ice at the same time. Furthermore, each dog has a fair chance of finding "holds".

c. Disadvantages. (1) The main disadvantage of the modified fan is that the tugs tend to get tangled; a good driver, however, generally can prevent this if he knows how to handle the long whip.

(2) Another drawback is that some power is lost in the sidewise pull of the flank dogs.

61. ESKIMO FAN HITCH. The true Eskimo fan hitch

Figure 24. Modified fan hitch.

Figure 25. Eskimo fan hitch.

(fig. 25) is the result of many generations of experiment and experience in the far North.

a. Description. In the Eskimo fan hitch, the lead line is about 10 or 12 feet long, and each of the single lines leading from the harness to the tie-in ring (attached to the lead line) is of a different length. The lengths vary from approximately 8 feet to about 14 feet. These lines are tied to the tie-in ring with a knot which can be untied easily with a quick jerk.

b. Advantages. (1) The tie-in ring at the end of the gangline, to which all the tugs are tied, gives a decided advantage over the tandem and double tandem hitches when quick release of the dogs is necessary in an emergency.

(2) The fact that each dog's lead line is of a different length gives this hitch an advantage over both the tandem and the other fan hitches, by permitting the dogs to fan out when the going is good, and conversely, to travel in a more compact and straight line than is permitted by the modified fan when the trail is bad and the load heavy.

(3) A further advantage in having the dogs at varying distances from the sledge is that no one dog can throw another off balance by a sudden lunge or jerk.

(4) This type of hitch, in common with all the fan hitches, permits each dog to work as an individual, picking his own way when the sledging is easy, but coordinating his efforts with those of his mates when load and trail conditions demand the utmost in cooperative effort.

c. Disadvantages. The disadvantages are largely the same as those applicable to the other fan hitches.

Section II. TRAIL OPERATIONS

62. GENERAL. a. The successful accomplishment of a dog-sledging mission is based on careful and complete preparation, and on a thorough understanding and anticipation of the nature of the mission. Equally important are an understanding of the conditions that may be encountered and a complete mastery of sledging technique.

b. Preparation. Before any sledging mission is undertaken, certain routine checks and preparations are essential. These may be logically carried out as follows:

(1) *Condition of team.* Make certain that the dogs are in good physical condition. Make sure you are using the right dogs for the mission. Check the condition of harness and sledges.

(2) *Condition of trail equipment.* Assemble and have ready all the equipment required for the mission, and check its condition carefully. Make certain there are no omissions. Check especially the condition and adequacy of tents, stoves and fuel, food boxes, trail marker flags, and extra clothing; but always keep in mind that the lighter the load, the better the going. The necessity for careful and adequate preparation cannot be too strongly emphasized; the success or failure of a mission may hinge on a single detail. Trail repair equipment can be kept in a tight box, placed on the sledge where it is readily available. Do not make the mistake of loading and lashing the payload on top of equipment which may be needed immediately while sledging.

(3) *Navigational preparation.* A briefing session is desirable. Adequate maps should be available. The route should be carefully selected and planned, and the possible types of terrain and conditions to be encountered should be considered. In rough terrain, a compass course (though always valuable to know) may be the most difficult route. Navigational aids should be checked for condition. Sending and receiving radio equipment should be carefully inspected to make sure it is in good working order.

63. LOADING. The most important element in loading is proper balance. Heavy equipment should be placed on the bottom of the sledge to avoid making the load top-heavy. The lower its center of gravity, the better. The heaviest weight should be placed just to the rear of the center; this allows easy maneuvering of the sledge on turns, with the gee-pole or the rear handle bars, and by the dogs. Lashing is an important consideration; unless the load is lashed tightly, it will keep shifting and lose its balance. The use of canvas tanks (par. 11) will simplify the securing of loads. Two ¼-inch ropes, 15 feet long, and rope loops between bridges should be used for lashing. One of the 15-foot rope lengths should be fastened to the front cross loop with a clove-hitch knot (fig. 26), and the other to a front or rear side loop between the

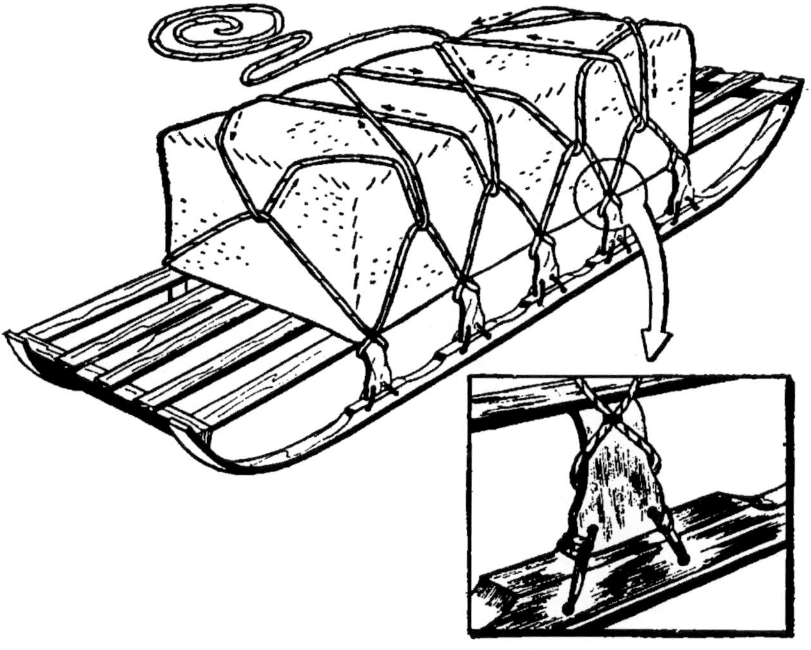

Figure 26. Lashing the load.

bridges. To lash a load properly, first fold the canvas tank flaps and then cross-thread each 15-foot rope through all the rope loops fastened between the bridges. Pull the ropes as tightly as possible and tie them. Unless very light, the load should not be more than 2 feet high. A high load has a decided tendency to cause swaying, increases the danger of overturning, and makes the sledge exceedingly difficult to maneuver.

64. WEIGHT OF LOADS. The average load should be calculated on the assumption that each dog will pull his own weight. However, loads vary widely with conditions and terrain. On flat terrain, where normal sledging conditions may be expected (as on the Greenland Ice Cap), a load of from 1000 to 2000 pounds is not exceptional for a team of nine to ten dogs. But the maximum load should not normally exceed 600 pounds. In rough, mountainous terrain (as in the Alaskan and Greenland mountains), about 300 pounds is a good average load and 400 pounds should be the maximum *at any time.* Heavier loads are very difficult to control in hilly or rough areas. Moreover, not only is excessive weight hard on the sledges, but in crevassed areas it is dangerous as well, because it puts too great a strain on the snow bridges over crevasses. When loads exceed 400 pounds, a part of the load should be carried on a trailer sledge. The trailer sledge should have the lighter load.

65. LEADING. a. In sledging operations involving heavy loads, it may be good practice to have a man travel ahead of the first team, either on skis or on snowshoes. Both Peary and Amundsen used this method effectively in sledging to the North and South poles, respectively. (With light loads, of course, travel would be too fast for a man to stay ahead.) A fourfold purpose is accomplished by having a man out in front:

(1) The dogs usually pull better.

(2) A more direct course is followed (very few lead dogs can ran as true a course as a man).

(3) The man in front is better able to pick a good trail.

(4) The trail is partially broken.

b. At the start of the day's sledging, when the dogs are usually high-spirited, it may be necessary to permit the lead dog to break the trail, for even with heavy loads it will be quite difficult for a man to stay ahead. However, if an extra man is available, he should start out immediately after breakfast and break the trail—at least until the teams catch up with him. The type of terrain will also determine whether a man should go on ahead of the teams. Sea-ice, such as is found in the Greenland and Baffin Island areas, often makes for fast sledging which would make it too difficult for a man to keep ahead; sea-ice, furthermore, is sometimes rough and almost impossible to ski over. Riding on the sledge will usually be necessary under such conditions.

66. ARRANGEMENT OF TEAMS. Teams should travel in single file so that all use the same trail. At no time should teams be so far apart as to lose sight of each other. An emergency may arise which will require the immediate assistance of another driver. This is especially true in treacherous crevassed regions. The best teams should be used in the lead to break the trail. If the going is difficult, change lead teams every hour or two. Slower teams hauling heavier loads, following in the rear, will have the advantage of a broken trail. In addition, the dogs traveling behind will work better than the front dogs, because they will attempt to catch up to those ahead of them.

67. RATE AND DISTANCE OF TRAVEL. Lightly loaded mail teams in Alaska often average 60 miles a dog. With heavy loads (100 pounds per dog), a 20- to 25-mile daily average is considered good. The condition of the surface will be the limiting factor, but with a good sledging surface, an average of three to four miles an hour with a heavy load is possible. On long sledging trips the day's sledging distances should be limited, in order to conserve the dogs. In warm weather, when there has been a good deal of thawing during the day, it is generally best to travel in the evening or at night. Snow becomes crusted over at night because of the lower temperatures, and thus provides a much better sledging surface.

68. MAINTAINING RATE. Often, when traveling over long distances in barren country, the dogs will become disinterested, bored, and lax in pulling. This will necessitate "pepping up" the dogs at frequent intervals. Snapping the whip, calling individual dogs by name, or talking to them helps to produce renewed alertness and energy. Sometimes a bell on the lead dog will relieve the monotony. Some drivers find that blowing a whistle or ringing a bell is helpful when the team lags. One famous driver in the "All-Alaska Sweepstake" race carried a bell during the run. Before the start of the race, after hitching the dogs, he lightly tapped each dog on the rump with his short whip, and at the same time rang his bell. When he wanted more speed during the race, he rang the bell. The association of the bell with the blow on the rump caused the dogs to speed up.

69. RELAYS. Sledging conditions may be such as to require relaying the loads. If a trailer sledge is being used, it may be unhitched and the front sledge may be taken on ahead. When traveling through soft snow on a windy day, it is not advisable to sledge more than a mile without going back for the trailer sledge; otherwise it may be impossible to use the old track before it drifts over. After the front sledge has been brought forward, unhitch it and make the return trip by ski-joring behind the entire team (par 48b (3). As a safety measure against sudden storms or darkness, which might result in the driver's getting lost while making a relay, colored marker flags should be placed at short intervals on the trail. These may be put out on the trip with the front sledge and picked up on the trip back with the trailer sledge.

70. REST PERIODS. Opinion varies as to the rest periods to be allowed dogs on the trail. The decision is often governed by the endurance of the man skiing in front of the team. The best advice is to know and watch your team and to use your own good judgment. On long trips, with heavy loads, it is common practice to rest 5 or 10 minutes every hour. With a good surface and light loads, the dogs can be run for 3 hours or more without a rest. Travel over hilly or difficult country requires frequent stops. Some teams are extremely difficult to start when tired, but, once under way, they usually keep going. It is best to run such teams for long periods without rests.

Take advantage of days when traveling conditions are good to help conserve the dogs. Base the schedule either on sledging time or on distances traveled, depending on conditions.

71. RIDING ON SLEDGE. Riding is not advisable at any time, unless the sledge is lightly loaded and the surface is good. Riding on the sledge may be necessary when the journey is over very rough sea or pressure ice, but the added weight of a man will reduce the rate of travel. Normal driver travel should be done on skis, on snowshoes, on foot, or possibly on a "ouija board" (par. 18, fig. 22). If the driver is on skis, he may get some help during rapid travel by holding on to the gee-pole, or the rear handle bars, or to a tow-rope attached to a sledge bridge. To keep the dogs in the habit of pulling, never allow them to haul a completely empty sledge. It is sometimes an excellent procedure for the driver to sit on the sledge and remain seated even on an upgrade. From this position he can help the dogs merely by pushing with one leg.

72. STARTING HEAVILY LOADED SLEDGE. a. After a stop, sledge runners occasionally freeze slightly to the surface. With a heavy load, this may make starting extremely difficult. The main objective in starting is to get each dog to "hit" his harness simultaneously on the initial pull. This is accomplished by standing at the side of the gee-pole and pulling back on the towline to tighten all the tug ropes (fig. 27). With a command such as "All right", there will be a concentrated pull on the sledge. At the instant the command is given, the driver should move the gee-pole sideward to help break the runners loose and put the sledge in motion. Lacking a gee-pole, attach a short rope to the front or back of the sledge, and pull it sidewise. If a basket sledge or a Greenland sledge is being driven, rear handlebars are used.

b. Trailer sledge. A trailer sledge places considerable strain on the dogs in starting. When the sledges are stopped, the driver should pull upward on the trailer

Figure 27. Starting heavily loaded sledge.

hitch attaching the two sledges (fig. 28). This will pull the lighter trailer sledge forward about 1 or 2 feet. It will then be started by the jerk of the front sledge when it gets into motion. Thus, most of the starting strain will come only from the weight of the front sledge. If the nose of the rear sledge is pulled slightly to one side, the pull of the dogs in starting will break the runners loose if they are sticking to the snow.

73. ICING SLEDGE RUNNERS. a. To reduce surface friction in the komatik-type sledge, the Eskimos of the western Canadian arctic commonly freeze at 1- to 2½-inch coating of fine grit-free mud—applied fairly wet—around and up the sides of the heavy runners. The mud is held on mainly by freezing, but adhesion may be partly aided by driving small nails 4 to 6 inches apart, and at an upward angle, into each side of the runner. Mud is then placed on the runners, and after freezing, is planed, filed down, or scraped with a knife to a smooth surface. Water is next frozen over the mud. The coating should be about 1/8 inch thick (the colder the weather the thinner

Figure 28. Starting trailer sledge.

the coat). Eskimos squirt the water on a bearskin with their mouths and then pass the skin over the frozen mud on the runners. Occasionally pieces of the mud coating may chip off and must be replaced. This method can be used only during cold weather, and only on solid runners. Resilient runners crack the mud quickly. Cooked oatmeal or flour is regarded as an even better material for this purpose than mud.

b. Eskimos of the eastern Canadian arctic sometimes use mud to widen runners as much as 10 inches, to prevent their breaking through crusted snow. Loads of 3,000 to 4,000 pounds have been placed on a 23-foot sledge and drawn by a 15-dog team, with runners treated in this way. It is estimated that at 50° below zero, the surface resistance of a sledge shod in this manner is only a quarter as much as with untreated steel runners.

74. EFFECT OF COLD ON RUNNERS. Extreme cold temperatures form hard, sharp angles on grains of snow, which then have somewhat the effect of sand on sledge runners. The use of steel runners on this type of surface more than doubles the normal strain on the dogs. Since wood provides a better running surface under these conditions, it is advisable to place wooden "shoes" over the steel-shod runners.

75. USE OF THE GEE-POLE. If heavy loads are to be hauled, any sledge, except the trailer type, should be equipped with a gee-pole or rear handlebars. They are invaluable for steering and for helping to start heavy loads. The gee-pole is useful also for holding the sledge back when going downhill. In addition, it may be grasped and used as an aid in skiing alongside the sledge. This puts the driver closer to his team—an advantage both for control and for working efficiency.

76. BRAKING THE SLEDGE. a. Nansen or Antarctic freight type sledges are not equipped with mechanical brakes. With these sledges it is necessary to use rope brakes in addition to the gee-pole. Roughlocks should be used on steep hills (par. 9f and fig. 31). Choke chains may be used but they are hard on wood runners. Basket sledges have a mechanical brake which can be applied by foot pressure while the sledge is in motion.

b. Dogs may be trained so that, when hitched in fan formation, a whistle will make them slow down and divide into two groups, one on either side of the sledge; the sledge then runs over the traces and gets ahead of the dogs. The driver pulls the handlebars back, tipping the sledge backwards, and lets his feet drag in the snow between the runners. Meanwhile the loaded sledge drags downhill (fig. 29). On rocky ground, the traces or tug

Figure 29. Braking sledge with dogs.

ropes should be lifted over the handle bars. If the sledge is too long, or if another type of sledge is used, the single tug or the tugs can be fastened to the rear of the sledge. While this method is widely used by the natives in Greenland, it must not be used with dogs not accustomed to it nor by a driver who does not have considerable experience.

77. TRAVEL IN EXTREME COLD. a. Lung frosting. In temperatures of 50° below zero or lower, it is unwise to run dogs or make them pull extremely heavy loads over long periods, because of the possibility of lung frosting. Such frosting results from heavy, fast breathing at these low temperatures. In the Arctic, sledge dogs have been lost as a result of this condition.

b. Drying dogs' feet. During cold weather it is imperative that water be kept off the dogs' feet, because ice may form into balls and crack the skin between the toes. When a dog's feet become wet, as may happen in crossing a stream or breaking through a thaw pool, dry them by running the dog in soft snow which acts as a blotter and takes most of the water off the feet. Immediately afterward, be sure to wipe off any remaining moisture. If balls of ice are found frozen between the toes, they must be removed. Usually the dog will chew them off, but occasionally the driver will have to remove them.

78. TRAVEL IN SPRING. Spring in arctic regions is marked by wide fluctuations of temperature during a 24-hour period. At this time, when there are long periods of daylight, it is advisable to sledge between sunset and sunrise when it is cooler and the snow offers less friction to the runners. The increased warmth of the sun at this season causes snow and ice surfaces to become sharply serrated. This surface condition is very hard on the dogs' feet and will result in cut and bleeding pads after very little travel. For protection of its feet a dog must then be fitted with boots of canvas or leather (par. 15, figs. 15 and 16). Always be sure to take the dogs' boots off for the night.

79. NAVIGATION. a. Trail navigation. (1) The methods of navigating with dog teams on the trail are essentially the same as those used by marine navigators. The sledge driver who has familiarized himself thoroughly with the contents of any standard text on marine navigation, such as Bowditch, Dutton, or Wylie, should be well-equipped for the responsibility of guiding his party to and from a desired objective.

(2) When traveling considerable distances in unfamiliar regions, the navigator should be able to check his dead-reckonoing positions by celestial observations. However, it is unlikely that dog-sledge operations in connection with military activities in the arctic, and even those involving rescue operations, will require skilled celestial navigators. The main emphasis should be placed upon dead reckoning, which is the method of keeping track or account of the courses steered, together with distances covered from the last known location. The first essentials are therefore the latest charts of the region to be traversed and a log book in which careful, systematic records of the courses and distances, together with other pertinent data, are kept.

(3) The two indispensable instruments for dead reckoning are a compass and some sort of device for measuring distances. If the route lies over land where there are few or no marks toward which a course can be set, a floating dial boat compass mounted on gymbals, or an airplane compass, is the best type. If on the other hand, landmarks are common along the trail to be covered, a simple prismatic or other type of small direct-view compass can be used. The navigator then takes a sight on some landmark on the route from a known position, and plots this direction as the course on the map.

(4) If circumstances make it necessary to steer by the compass without the aid of landmarks, the boat or airplane compass should be mounted at either the front or the back of the sledge, depending on where it can most easily be seen by the navigator as he travels alongside or behind the sledge. He can then keep constant watch on the course and call out to the man leading the way in

front of the lead dog "Right", "Left", or "Steady", whichever is necessary to keep on the course. Care should be taken to see that the lubber line of the compass is mounted exactly parallel to the axis of the sledge—that is, over a line which would bisect the sledge longitudinally into two equal parts.

(5) All metals which might affect the compass should be removed from the navigating sledge. If a completely nonmagnetic sledge can be arranged, one of the usual troublesome compass errors, occasioned by deviation, can be eliminated. It will then be necessary to apply but one correction to the magnetic compass reading, the error resulting from variation. This error springs from the simple fact that the geographic and magnetic poles are more than 1,000 miles apart. The compass needle points to the magnetic pole, whereas the desired direction is true north, or the direction of the north geographical pole. Therefore, except in those rare cases where the magnetic and geographic meridians coincide, the compass needle will be deflected either to east or west of true north. All courses should be plotted on the charts as true. If the navigator will remember this single simple rule he will have no difficulty: *M* to *T* add *E*. *M* stands for magnetic reading; *T*, for true north; and *E*, for easterly error. In other words, when converting the magnetic compass reading to true, the easterly variation is added; conversely, if the variation is westerly, it is subtracted. Thus if the magnetic compass reading is 37° and the variation is 8° easterly, the true reading will be 45°; if the variation is 8° westerly, the true reading will be 29°.

(6) A simple type of sledgemeter (fig. 30), and one which has long been used by polar explorers, consists of a bicycle wheel to which a cyclometer or measuring device, calibrated in geographic or nautical miles, has been adapted. This is attached to the rear of the sledge and is so hinged between supports that it can ride over irregularities of the terrain. Surprisingly accurate results can be obtained by this means.

(7) It is always *advisable* to mark the trail if it is to be retraced on the homeward journey. If the route lies

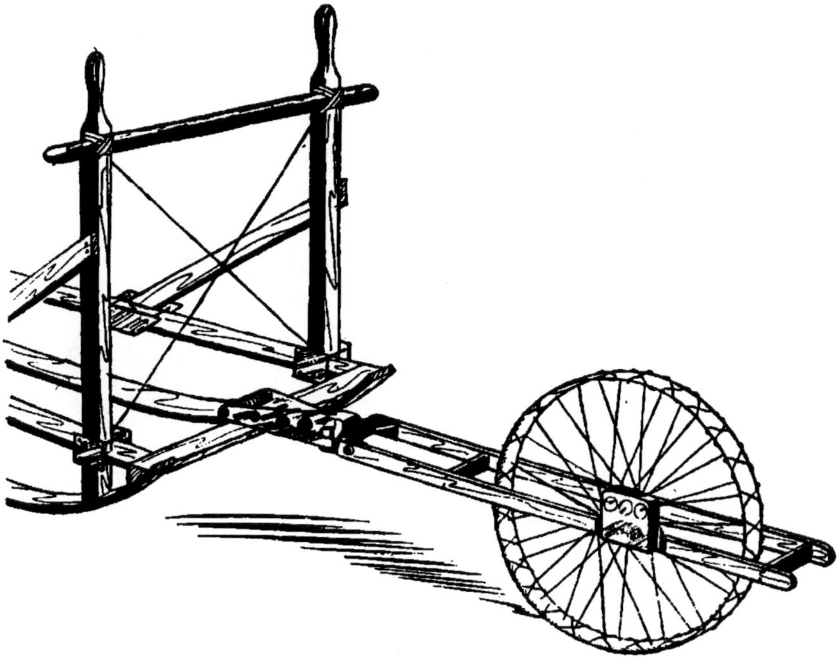

Figure 30. Sledgemeter.

across snow and ice or other terrain where there are no landmarks, marking the trail becomes a "must" in the interest of safety. Carry small flags, about 1 by 1½ feet, mounted on thin pieces of bamboo or other material, about four feet long, which will provide light, easy-to-carry, durable markers. Tests have shown that the best color, from the standpoint of visibility against a background of snow-covered terrain or lands covered with green vegetation, is a bright orange or "flame scarlet". However, experience has shown that any similar shade will be found quite satisfactory. A recently developed material which is being used successfully is described as "fluorescent satin." It is a bright rose or cerise.

(8) In addition to the equipment described above, the navigator's kit should contain dividers, scale protractor, and a good supply of hard pencils.

(9) If lost during daylight, seek the highest lookout in the neighborhood and try to spot a known landmark. When night comes, build a snow house or pitch tent.

Think back over the route, and make an effort to determine mistakes in direction. Until certain of the way, it is futile to keep going.

b. Airplane reconnaissance. Prior reconnaissance by air is always of great assistance in laying out prospective courses for sledging operations. If aerial photographs of the area to be traversed can be provided, a considerable part of the work of navigating the party is at once eliminated.

80. SLEDGING TECHNIQUES UNDER SPECIAL CONDITIONS. a. General. Wide variations in terrain, sea-ice surfaces, glaciers, and weather conditions, pose varied and often unpredictable sledging problems. There is no one standard for travel. Similarly, no hard and fast rules can be laid down for equipment or techniques. For example, various types of hitches have been described (pars. 56–61) and each of these has its own use and application under certain conditions. Familiarity with all types of hitches and what to expect of each is desirable.

b. Land travel. (1) The greatest problems involved in sledging on land surfaces are encountered in mountainous country and in thickly wooded regions such as are found in Alaska or the Yukon. In negotiating hills, it is better to overbrake than to underbrake, for once a sledge gets out of control serious damage to both dogs and equipment may result. In hilly country the double tandem hitch is the most efficient. The single tandem is sometimes required in wooded areas. It is especially difficult to navigate in bushy, wooded country, for in such country compass courses are difficult to follow and readings must be taken at frequent intervals (par. 79).

(2) In *deep or fresh snow* it will often be necessary to break trail. Skis can be used for this purpose though snowshoes are preferable. The width of the trail will vary between 18 and 25 inches. The man who breaks trail should be relieved from time to time. If the going is very heavy and several teams are on the trail, the first team should be relieved now and again. The best type of

sledge to use under these conditions is the toboggan. The toboggan should have a gee-pole and rear handle bars which are long enough to permit pushing while wearing skis or snowshoes. The single tandem hitch is advisable for travel on deep or fresh snow in wooded country, although the double tandem hitch may be used. Sometimes the man in front will have to pull with the dogs. He should be about 15 feet ahead of the lead dog. A suspenderlike harness for this man can be contrived of ropes; the bow-line on the bight is the best knot for such a harness. If a rucksack is available, attach the trace to the lower part of the frame.

(3) On the *Greenland Ice Cap* where loads are generally heavy (in the case of self-supporting parties), a long ski-runner-type sledge (Nansen or Antarctic) should be used. The type of hitch to be used at any one time will depend upon the condition of the surface.

(4) In *rough, mountainous country* sledging is very difficult. The technique of sledging in such regions is as much of an art and a sport as skiing. If the country is timbered, the sledge generally used is a basket-type sledge with hinge brakes, slew brakes and large ski-type runners. Whatever the type of sledge used, it should have rear handle bars. Gee-poles, though useful as a rule, will not serve in timbered country. The single tandem hitch is not recommended for mountainous terrain because it is too long. The double tandem hitch is best here. Fan formations must be avoided except on glaciers, d below.

(5) *Downhill sledging* necessitates brakes or roughlocks (par. 9). Sometimes it may be necessary to wind a rope around the runners (fig. 31.) This procedure should be avoided as much as possible as it causes delays. Almost every kind of hill can be traveled with ordinary roughlocks, a man at the brakes, or a man at the gee-pole, using his skis or snowshoes as brakes (fig. 31). Also, the dogs can be fastened to the rear of the sledge. For very steep hills, it may be necessary to lower dogs and sledges separately on ropes. To belay dogs and sledges, push a ski-pole (or ice-ax) into the snow—handle first—at the top of the hill. Then make a dead turn around the ski-pole with

the rope. The man holding the other end of the rope should dig his heels solidly into the snow.

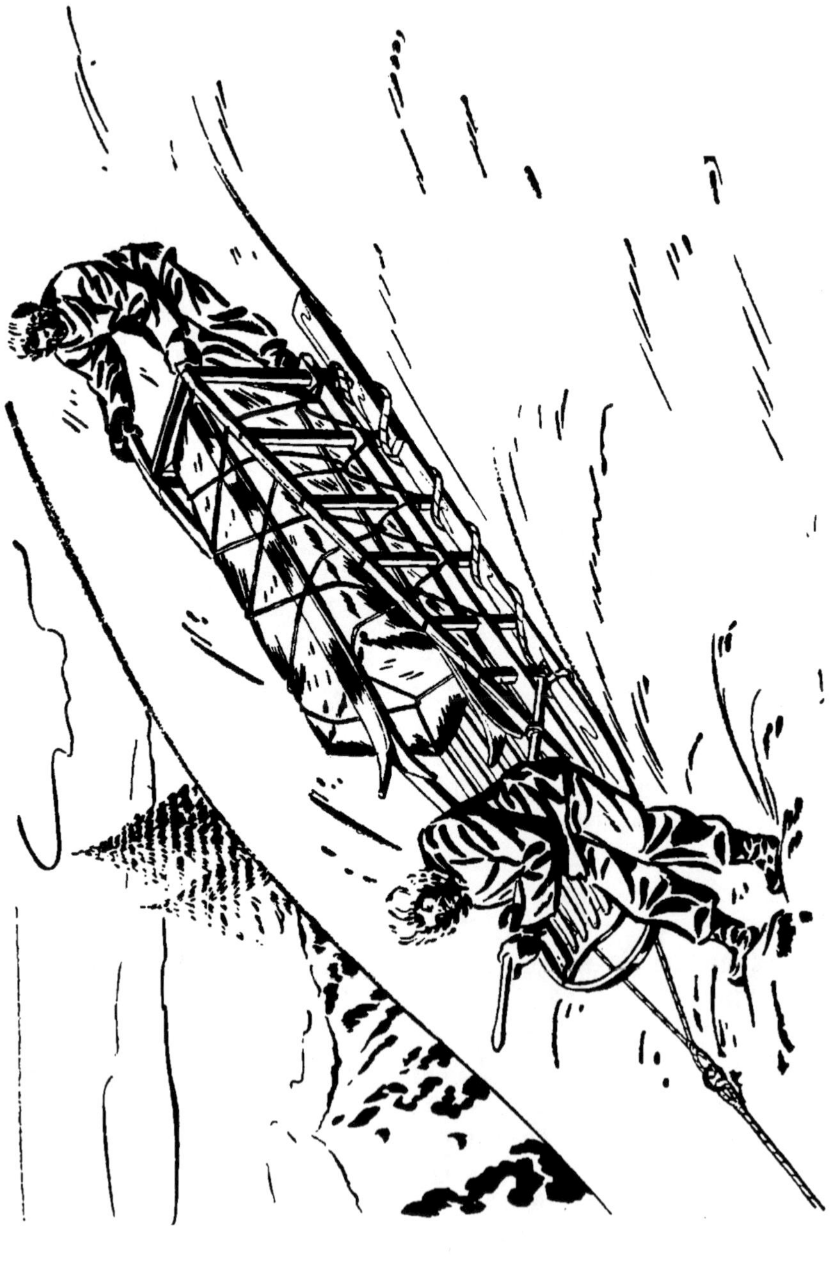

Figure 31. Braking sledge on steep hill.

① *Standing on uphill runner. (Bridle rope arrangement shown in insert.)*

Figure 32. Sidehill sledging techniques.

(6) Occasionally, *sidehill* sledging may be necessary. This is always difficult and sometimes dangerous. Good slew-brakes (par. 9) reduce the danger of sideslipping to a minimum. The front part of the sledge may tend to slide downhill. To offset this tendency, shorten the downhill part of the bridle rope by tying it to the side of the front bow. Smooth hills can be crossed simply by standing on the uphill runner of the sledge (fig. 32①), although the weight of a man on the runner places considerable extra strain on the dogs. Steep slopes require the use of a rope attached to the sledge. One of the men on skis pulls the rope from the upper side. Sometimes the second man will have to push upward against the sledge from the lower side and drive the dogs upward at an angle to the line of travel. If an Antarctic sledge is used, the best type of control is achieved by two men holding to a rope at each end of the sledge and pulling upward from

② *Holding with ropes.*

Figure 32. Sidehill sledging techniques — Continued.

the upper side. To obtain maximum leverage for the prevention of tipping, the ropes should be fastened to a bridge on the downhill side. The ropes are swung over the top of the sledge load. Consequently, the pull is from the top and at a higher and more natural angle to the puller (fig. 32②).

(7) In *downhill timber,* well-trained dogs hitched to a basket sledge can usually pass through the densest spots. The superstructure of the sledge brushes the branches aside. The driver protects himself by squatting on the runners behind the rear handle bars. He must be careful to place his hands on the upper crosspiece of the handle bars and not on the uprights.

c. Travel on sea-ice. (1) For safe travel, sea-ice should be 8 inches or more thick. Sea-ice less than 6 inches thick will rarely support a man's weight. Traveling on sea-ice always involves some danger. Ocean swells, currents, or sudden gales may shatter even the heaviest ice. Good sledging ice is never continuous for very long. As a rule, there are more numerous smooth stretches on landfast ice than on pack-ice, but traveling only on landfast ice is not always possible.

(2) There is a variety of opinion as to the best type of sledge for travel on sea-ice. The same is true as to choice of hitches. Although the double tandem hitch has been successfully used on sea-ice surfaces, one of the fan hitches seems preferable. It can hardly be questioned that the fan hitch is more suitable for crossing thin ice or narrow leads. If there is a break-through, the fan hitch gives the dogs greater freedom to choose their way. The team can easily spread out, thus achieving better distribution of weight. It is not always sound to argue that in case of a break-through the double tandem allows some of the dogs to get safely across, thus enabling them to pull the others over; it is entirely possible that the dogs who have broken through may also haul down those who have weathered the crossing. Whatever the type or condition of the ice, always be on the alert. Danger may appear when it is least expected; nothing is more treacherous than ice that looks safe. When sledging on sea-ice, wear

Figure 33. Sledging on thin sea-ice.

skis with broad straps for bindings; such skis can be put on and taken off as easily as slippers.

(3) The important thing in sledging on *young ice* is the distribution of weight. Therefore, a type of sledge with long, wide runners (Nansen, Antarctic, or basket sledge) is best on such a surface. A fan hitch, with an extra 5- or 10-foot section of towline between the bridle and the wheel dogs, is preferable to other hitches for travel on young ice. One man should ski far ahead. The others in the party are towed on skis behind the sledge (fig. 33), holding on to a rope about 10 or 15 feet long. Speed is imperative. If the double tandem hitch is used, it is advisable to unsnap the necklines, and it may be necessary to relay loads to decrease the weight of the sledge. It may not be wise for a team to use the track made by the advance man because of the cracking of the ice. The man traveling in front of the team, preferably on skis or on snowshoes—never on foot—should test the ice frequently with an ice chisel. Although the chisel may go through the ice (fresh or salt water) at the first blow, the surface may still be sufficiently strong to support the sledge. Experience is the best teacher in determining when ice is strong enough and when it is not.

(4) A solid, heavy sledge, such as the komatik (Greenland sledge), is the best type for travel over *rough ice or ice with pressure ridges.* Of importance are the rear handle bars, which are used to push, pull, lift, or straighten the sledge. In the matter of hitches, the fan hitch is preferable to others for rough ice, although the double tandem may also be good because there will be less chance of the shorter tug ropes becoming snagged on jagged ice. The fan gives the dogs freedom to jump over jagged ridges, to choose their own way or, if necessary, to come back to close formation. It may sometimes be necessary to lift the dogs, one by one, over ridges, and for this procedure the true fan hitch is best.

(5) In winter the roughness of sea-ice may be leveled off by a heavy covering of snow. With wind, this snow-covering is formed into ridges and furrows—sometimes several feet high—in the direction of the prevailing winds.

These ridges resemble the so-called "sastrugi" observed on the Greenland Ice Cap and in the Antarctic. The ridges should be chosen for travel, since they are composed of hard snow. Great caution should be exercised in traveling on snow-covered sea-ice. It is impossible to tell newly frozen leads covered with snow from solid old ice. Thin ice is indicated by gray spots on an otherwise snow-sovered surface. Thin spots should be tested with an ice chisel before crossing.

(6) In spring it is much easier to sledge after sunset than during the day. After the sun goes down there is a sharp drop in temperature; hence the surface—soft and wet during the day—becomes hard. On a *spring-ice* surface, the fan hitch has advantages over other hitches. The driver should be on foot and stationed at the rear handle bars of the sledge in order to steer it quickly and effectively between pools of water or holes in the surface. Somtimes it is possible to cross ice which is covered with water up to 10 inches. A narrow open lead can be bridged by filling it with sufficient chopped ice to insure buoyancy that will bear the sledge. There may be times when it will be necessary to use a skin or canvas sledge boat. Provided the load is not too heavy, any canvas which wraps the whole sledge should serve. Leads which have to be crossed with a sledge-boat may be covered partly or wholly with very thin young ice.

(7) Sledging on detached, floating floes (*pack-ice*) with open leads and pools between them calls for sheer acrobatics. It is impossible to make helpful suggestions about navigating pack-ice. If driver and dogs know each other, they will come through intact. Dogs are peculiarly sensitive to danger and will not fail in a "tight spot." When sledging on pack-ice, never wear skis or snowshoes—life depends on agility. Cross open water with a canvas sledge-boat or use a floe as a boat. Also, with experienced drivers and good dogs, the dogs may be made to serve as a tug. Hitch them to a floe in fan formation, and let them swim for it.

d. Travel on glaciers. Negotiating crevasses in glaciers or shelf ice is the most dangerous of all dog-sledging operations. Sledging over crevasses has cost the lives of a number of competent Arctic and Antarctic explorers. It is useless to attempt to establish any hard and fast rules for this extremely hazardous work, for no two situations are exactly alike. Extraordinary judgment and a capacity to meet emergencies will be required. The following suggestions will be useful:

(1) If aerial photographs are available, study the crevassed area to determine the best possible route. The examination of photo pairs under a stereoscope will provide a detailed and accurate picture of the terrain and is highly desirable.

(2) If possible, make a preliminary survey of the area to be traversed. This is best accomplished by a party of three men on skis, tied together with a rope at 30- to 40-foot intervals, and each carrying an ice-axe or pole. The best knots for the tying are the bowline for the two end loops, and the middleman loop for the middle one.

(3) Never walk on snow-covered glacier ice unless you are wearing skis or snowshoes and are roped. This should be a standing order for all sledging operations.

(4) Riding a sledge over a crevasse is very dangerous and should be avoided whenever possible. Crossing a crevasse on skis or snowshoes which distribute the weight is safer than riding the sledge; however, riding the sledge is safer than walking *without* skis or snowshoes for the weight is better distributed. If an injured person must be transported over a crevasse, he will of course have to be carried on the sledge; in this event, however, every effort should be made to avoid any other weight on the sledge.

(5) The double tandem hitch, the modified fan hitch or the Eskimo fan, is recommended for travel over crevassed areas.

(6) Be sure to distribute weight as evenly as possible over all the sledges. Furthermore, see that the food and essential equipment are so distributed among the sledges that each is a complete unit; thus, if one sledge should be lost in a crevasse, the whole party will not be stranded.

(7) Extreme caution must be observed at all times in

crevassed areas, but speed is essential when a crevasse is actually crossed. Never sledge over the bridge of a crevasse until a trial crossing has been made first by a roped man on skis. Having a man ahead of the team, and across the crevasse, will also tend to make the dogs pull faster.

(8) When crossing the most dangerous areas, many sledgers prefer not to risk a life by having a man ski out in front of the team—unless he is on a rope tied to the front of the towline and on skis or snowshoes. In this case, once across the crevasse he should release the rope, so that if the dogs break through they will not pull him backward.

(9) The driver should never cross the snow bridge over a crevasse from his normal traveling position alongside the sledge. This will be the most likely place or combination of conditions for a break-through because of the concentrated weight of the man and the sledge. When crossing a dangerous bridge, a 30- to 40-foot rope should be attached to the rear of the sledge. The driver can use this rope to pull the sledge aside, so that the sledge will not cross the crevasse in the same place already weakened by the crossing of the dogs. Finally, the driver can use the rope to let himself be towed across the crevasse roof, pulling himself still further off to one side to avoid the trail made by the dogs and the sledge (fig. 34). It is generally safest to cross a crevasse at right angles.

e. Use of skis and snowshoes. During sledging, skis and snowshoes must be used almost continuously, except where rough ice or snow conditions make it impossible to travel, or when it is necessary to run behind the sledge as, for example, on broken pack-ice. Skis and snowshoes each have certain definite advantages. Generally speaking, skis are most useful, except in deep powder snow or densely wooded country. Even persons with little experience can usually travel more easily and quickly on skis than on snowshoes. Speed is, of course, an important factor in sledging, especially in crossing crevasses. Keeping up with the sledge on snowshoes is difficult when the pace is fast, because you cannot help yourself along by

Figure 34. Crossing crevasse.

hanging onto the gee-pole, the rear handle bars or a tow rope. However, if the party is reduced to man-hauling the sledges, snowshoes will generally prove more effective than skis.

81. RESCUE WORK AND AIRPLANE COOPERATION. **a.** On rescue missions, full advantage should be taken of sledge and airplane cooperation. Speed is vital in rescue work and may be greatly enhanced by supporting sledging operations on the ground with planes. Two-way radio communication will facilitate cooperation between the sledge team and the plane.

b. The reconnaissance advantages offered by the use of an airplane should be fully utilized. A survey of the rescue route to be traversed by sledge may be carried out in a matter of an hour or two by plane and will serve to point out the safest, quickest, and most desirable route.

c. Airplanes may be used to transport sledge teams speedily to landing areas as close as possible to the scene of disaster (par. 83d). While suitable landing sites may not be available in the immediate vicinity, the closest possible approach in the shortest possible time may save lives.

d. The lighter the load a sledge carries, the more rapidly it will travel. Through airplane support, loads may be reduced to a minimum. If permitted by weather conditions, the dropping of supplies from aircraft is the most efficient method of supplying ground parties. A complete account of such operations is given in FM 31–40. As far as dog sledging is concerned, the responsibility of the ground party is twofold: first, making known the needs of the sledging party; second, identifying the ground party for the supplying aircraft. The former will normally be foreseen by the sledging party which should prepare a list of anticipated needs to be turned over to the supplying air unit before departure from base. Identification is most easily effected by means of smoke bombs or by panels or strips of orange or "fluorescent" cloth laid out according to prearranged patterns to indicate certain messages. Smoke-bombs provide the best method of attracting the attention of the aircraft, because not only can smoke be seen for greater distances than cloth panels, but it also provides the pilot with a more accurate knowledge of the wind direction. The best use of panels is to mark the target once the aircraft has located the party. Select a spot as free as possible of obstacles dangerous to the maneuvering of aircraft. Avoid crevassed areas, or, if this is impossible, try to pick a crevasse-free area at least 100 yards square. The plane usually approaches the target against the wind. Keep dogs and personnel well to one side of the course. The only exception to this rule is when there is a strong wind which might carry the wind-filled parachutes out of reach or even into crevasses. Under such conditions, the personnel of the party should stand by to collapse the parachutes as soon as they reach ground. The men should be spread out and each man should make a mental note of the number of packages dropped and where they land.

82. MAKING CAMP AFTER DAY'S WORK. a. Camp should be made at a place where the dogs will be sheltered from the wind, but not close under a bank where a snowslide or drift may cover them. Shelter from rain is also advisable, as dogs soon become exhausted and chilled by exposure to dampness.

b. Care of the dogs should be the first concern of the driver after each day's sledging. Each dog should be inspected for sore feet and any other injuries and given necessary first-aid treatment.

c. Dogs should be fed only once a day, at the end of the sledging period. They should not be fed before or during work, for this will result in listlessness and vomiting if there is a heavy load and hard pulling. Most trail dog food is put up in 1¼- to 1½-pound cakes which are adequate for normal-size sledge dogs. If heavier dogs are used with lighter ones, allowance for the difference in weight should be made in the food ration.

CHAPTER 6

TRANSPORTATION OF DOG TEAMS

---

83. GENERAL. Every means of transportation should be utilized to move dogs as close as possible to the area in which they will operate. The types of transportation used will depend on the terrain to be traversed and on available transport facilities.

a. Movement by truck. (1) When suitable roads are available, dogs can be efficiently transported over long distances in trucks. A stake body truck will serve the purpose well. Each dog should be put in an individual shipping-crate kennel (fig. 19), which will eliminate fighting. The crate itself can be used at the base destination as a kennel. These shipping crate kennels may be double-decked in two rows, arranged back to back on the truck, but they must be securely lashed. At least sixteen crates can be transported in this manner on a 1½-ton truck. The number of dogs carried may be increased by placing a bitch and a male together in the same crate. Gear boxes and sledges may be lashed on top of the crate kennels.

(2) If short distances are to be traveled and the use of crate kennels is not advisable, each dog should be secured by a short chain to the side of the truck. With this method 15 to 20 dogs can be accommodated. If dogs are chained to the truck and are not in crates, an attendant should ride with them at all times.

b. Movement by rail. (1) Without equipment, as many as 60 dogs can be transported in an ordinary baggage car. With equipment, such as sledges and gear boxes, 40 dogs should be the maximum carried. The average boxcar will accommodate about half this number.

(2) The use of the combination shipping-crate kennel is advisable. The crates should be stacked in two tiers on one side of the car. Sledges and gear may be loaded on top of the crates. If crates are lacking, the dogs should be chained in three rows along the two sides of the car and down the center. Chains not over 18 inches long, and equipped with swivel snaps, should be fastened to the car floor with a small staple. Bitches should be interspersed among the males to prevent possible fighting.

(3) The dogs should be taken out of the crates before feeding. A tethering line may be fastened along one side of the car to secure dogs while they are being fed. It will also enable them to obtain a little exercise.

(4) An attendant must be present in the car at all times. A prepared commercial dog food is most convenient to use during transport by rail. Straw or shavings are desirable for use in keeping the car clean. *Proper ventilation must be provided at all times.*

c. Movement by water. (1) When transportation by water is necessary, dogs are best carried on the deck of the vessel. If the combination shipping crate and kennel with sliding door is being used, the dogs should be kept in the kennels during the night and during rough weather when the vessel rolls and the decks are awash. If possible, the crates should be raised several inches off the deck so that water will not wash inside them. All crates must be firmly lashed in place to prevent shifting when the vessel rolls. During calm weather, and in daytime, it is best to chain the dogs on deck. Do not secure them along the ship's rail where it may be possible for the dogs to go over the side; occasionally the vessel may roll and fling dogs overboard, or they may become excited, jump over the side and suffer strangulation.

(2) Water the animals at least twice a day, but feed them only once. A more or less continual clean-up of excrement is necessary, and there should be a general clean-up each morning. Decks should be hosed at this time. Particular care must be taken to prevent dogs from standing in salt water for long intervals as their pads may crack and become very sore. A watch should be

established at night; this is particularly important if the weather is rough and a large number of dogs are being transported.

d. Movement by air. (1) There may be occasions when considerable time can be saved in rescue operations by flying dog teams to the closest possible location in the vicinity of the forced landing. This has been accomplished successfully in Greenland and Alaska on several occasions. In Alaska, a B–17 has carried as many as 32 dogs (without equipment). The C–47, C–54, or similar cargo ships, are still better for dog transport. Two 9-dog teams with all their equipment may be transported readily in these types of planes. A tethering line should be fastened securely along each side of the plane in the after compartment, using neck lines no longer than 6 inches, if possible. This will allow very little room for the dogs to move about.

(2) Small aircraft which do not have sufficient space for tethering the dogs must sometimes be used where landing sites are uncertain or difficult. Where space is thus limited, the animals can be placed in light, compact crates. These crates should be as long as the dog and not over 15 inches wide and 15 inches deep. Such crates will eliminate thrashing if the dogs become excited. However, in most cases crating will not be necessary. The UC–64 (Norseman) would be the best of the small airplanes for dog transport. One team, without equipment, could probably be carried.

CHAPTER 7

SPECIFIC LOADS FOR DETACHMENT OF INFANTRY

---

84. GENERAL. An infantry detachment of 39 riflemen, one light machine-gun squad and one 60-mm mortar squad attached, together with field equipment, ammunition, and 4 days' rations for men and dogs, may undertake operations employing dog transportation in the following manner:

a. Train. Rations for 4 days for men and dogs, for a detachment of 39 riflemen, one light machine-gun squad, and one 60-mm mortar squad, together with kitchen equipment and tools, may be loaded on two sledges.

(1) Divide the equipment and supplies into two loads of equal weight. The field range, kitchen tentage, and tools should be kept together on one sledge.

(2) Next, place the field range on edge, just behind the center of gravity of the sledge. Articles should be placed around the field range to prevent it from shifting about.

(3) Crosscut saws, because of their length, should be placed on the bottom of the sledge with the teeth pointing inward.

(4) Axes should be sheathed and placed where they will be easily accessible.

(5) Other components of the load should be placed on the sledge, so that the heavier items are at the rear and the lighter ones toward the front.

(6) Small articles should be made up into convenient packages for loading. Empty gunny sacks are useful for this purpose.

(7) It is desirable that quarters of beef to be loaded be boned and put in light containers.

(8) Bread should be packed in a cardboard container or a canvas bag and loaded on the forward part of the sledge. Mattress covers make excellent bread carriers.

(9) Transportation of liquids and canned goods subject to damage by freezing should be avoided, and articles such as powdered milk and eggs should be substituted.

(10) Dog rations for the team should be loaded on the front of each sledge.

(11) For individual packs, see paragraph 86.

b. Combat loads. Two light machine guns and one 60-mm mortar, with a supply of ammunition, may be loaded on one sledge in the following manner:

(1) *Light machine guns.* Two light machine guns and 1,750 rounds of ammunition per gun should be loaded on a basket sledge as follows:

(*a*) Place a layer of ammunition chests, lying on their sides, on the bottom of the sledge.

(*b*) Place a light machine gun, muzzle to the front, bolt-handle up and its tripod folded, cradle to the front, on top of the ammunition chest at the rear of the sledge, gun on the left side and tripod on the right. Between trail legs and gun place an ammunition or spare parts chest.

(*c*) In front of the first gun and tripod place a second gun and tripod in similar manner but reverse the position and place the gun on the right side and tripod on the left.

(*d*) Place a light tarpaulin or canvas folded to appropriate size over the load, and lash everything securely to the sledge (if canvas tanks are not used).

(2) *60-mm mortar.* A 60-mm mortar, complete with 60 rounds of ammunition, may be loaded on the same sledge with the light machine guns in the following manner:

(*a*) In front of the light machine-gun load place the mortar and bipod on the right side of the sledge. On the left side place the base plate and accessories.

(*b*) In between these pieces and in front of them place the loaded ammunition containers.

(*c*) Cover everything with a canvas or light tarpaulin, and lash securely to the sledge.

Note. The regular sledge equipment and 4 days' rations for the dogs may be loaded in front of the two light machine guns, the 60-mm mortar, and their ammunition, and lashed with the rest of the load.

(3) *Ammunition.* The ammunition load of 160 rounds per man for a detachment of 39 men may be carried on one sledge in the following manner:

(*a*) Remove the metal containers with the ammunition from the wooden box.

(*b*) Place the four metal boxes of ammunition flat on their sides on the bottom of the sledge, starting from the center and loading toward the rear.

(*c*) The additional 240 rounds required may be put in a canvas bag and placed in front of the four metal boxes.

(*d*) Lash the load securely.

Note. The regular sledge equipment and rations for four days for the dogs pulling the sledge may be placed forward of the ammunition load.

85. WEIGHTS. The following table gives the weights of the articles comprising the above loads:

Load	Item	Pounds	Total
No. 1	1 range, field, No. 2	150	
	Tentage and tools	150	
	44 dog rations, 3 pounds each	132	
	70 rations, 5 pounds each	350	
			782
No. 2	44 dog rations, 3 pounds each	132	
	134 rations, 5 pounds each	670	
			802
No. 3	2 light machine guns	40	
	2 tripods, light machine guns	62	
	3,500 rounds (14 belts)	287 (389)	
	1 60-mm mortar	28	
	1 base plate and accessories	23	
	60 rounds, 60-mm ammunition	209 (260)	
	44 dog rations, 3 pounds each	132 (132)	
			781
No. 4	6,240 cartridge ball, cal. .30	450	
	36 dog rations, 3 pounds each	108	
			558

Note. Three sledges, with teams of eleven dogs each, should be used to transport loads Nos. 1, 2, and 3, and one sledge with a team of nine dogs should be used for load No. 4. Three extra dogs are provided for each team to pull the sledge driver and regular sledge equipment.

86. INDIVIDUAL PACKS. When troops are operating in very cold country, the special equipment required by the individual is decidedly heavy and bulky. For this reason it is desirable that whenever possible a portion of the soldiers' pack be transported by sledge. When the sledges can precede the marching columns to the new bivouac area, it will usually be possible because of their greater mobility for them to return empty and pick up the sleeping bags on a second load. The sleeping bags, packed with the men's extra clothing and weighing about 25 pounds each, can easily be carried by three sledges. By this method the packs may be lightened without increasing the amount of dog transportation, provided the march does not exceed nine miles.

87. HEAVY WEAPONS EQUIPMENT. Heavy-weapons sections may be transported by dog team in the following substitution of each type section for the rifle company weapons squads:

Load	Item	Pounds	Total
No. 3	2 heavy machine guns (cal. .30)..	67	
	2 tripods	102	
	5,750 rounds	471	
			646
No. 3	2 81-mm mortars (complete).....	272	
	60 rounds M–3 (light shell)......	600	
			872

Note. No. 4 sledge could be utilized to carry the second sections of a heavy-weapons platoon, or for extra ammunition.

CHAPTER 8

PACK DOG TRANSPORTATION

Section I. PACK DOGS

88. GENERAL. Dog packing is as old as sledge-team driving. The Indians and Eskimos of the far north use their sledge dogs in the summer to carry their belongings over the rough land and tundra in improvised pouch packs of hide and strapping.

89. ADVANTAGES OF PACK DOGS. Pack dogs are useful for both summer and winter operations; outposts in the far north find them very valuable as year-round aids to transportation. Trained sledge dogs are usually used in sledge teams during the winter; but the larger types of dogs are occasionally used in winter to pull individual pulkas (par. 91b), toboggans or sledges, as well as for packing in the summer. There is practically no place a pack dog cannot go with a load. As a rule, loads carried may equal half the dog's weight. Generally this amounts to about 35 pounds.

90. BREED OF DOG MOST DESIRABLE FOR PACKING. a. There is no definite clear-cut breed of dog for packing. The best packers are the heavier sledge dogs and the following breeds: Chesapeake Bay retriever, Great Pyrenee, Newfoundland, and Saint Bernard. The pack dogs are selected to meet certain standards. These standards are: weight over 80 pounds with dogs preferred from 110 to 135 pounds; broad heavy chest; heavy padded shoulders; no indication of sway-back condition; heavy hindquarters with straight hind legs. Dogs with these qualities are bred and used by natives of Northern

British Columbia, the Yukon Territory and Eastern Alaska, for packing and for heavy freighting by sledge and toboggan. Such dogs can survive hard trail work and extreme weather better than any other standard breed, and get along on less food. Indian dogs, which are generally of a mongrel nature, are often found with the same qualities. The Indian dog possesses exceptional durability and wiriness and, with the exception of the quality of the coat, they make as good packers as the better known breeds.

b. Use of pack dogs. The designation "pack dog" is misleading, since packing represents one of three jobs this type of dog can do.

(1) A dog of the above description can pack an average load of 35 pounds all day and every day over extremely rough terrain. The same dogs can pack 50 pounds for a limited time, that is, 1 or 2 days at a stretch. The dogs are always under the supervision of a handler and can make as many miles a day with loaded packs as it is possible for the most rugged man to hike. The packing is done when—

(*a*) The snow conditions are such that a sledge and toboggan cannot be adequately used (no snow or ice on which to travel).

(*b*) The timber is so dense that it prevents the use of toboggan or sledge with team.

(*c*) Less than a sledgeload of equipment is desired to be transported to a given place over snow trails.

(2) The groups of five or six pack dogs are equipped with harness, sledge, and toboggan, and are trained as a tandem freight team. When large, bulky loads that cannot be put in dog packs are to be transported, and terrain and weather conditions permit the use of sledges or toboggans, these pack-dog groups function as a team.

(3) Each pack dog is also supplied with one pulka complete, and is trained to pull a load on it by himself. Pulkas can be used where the density of timber prevents the use of a full team and larger sledge or toboggan, or where loads are small enough so that the use of a full team is not warranted. Over rough country a dog can

pull a load by pulka equivalent to his own weight. On level terrain he can pull three times his own weight.

91. EQUIPMENT. a. The pack. The pack now used is a refinement of the crude pack used by the Eskimos on their larger sledge dogs. The Eskimo uses two pouches of skin, with a fur strap across the dog's breast and a leather binding under the belly. This type of pack has a tendency to slip. The principles involved in dog packing are different from those involved in horse or mule packing, for a dog is loose-skinned. The belly-strap is therefore necessary, not as a cinch but as a stabilizer, to help prevent sidesway. The breast strap on the Army pack serves to stabilize the two pouches and to keep them from slipping back when the dog is walking up steep grades. The addition of a breeching prevents the pack from slipping forward.

b. The pulka. (1) *Description.* The pulka (fig. 35) is a short, basketlike toboggan made of wood veneer; it has an adjustable canvas cover. It is equipped with shafts, adaptable for use by both man and dog. It was first developed by the Laplanders who harness reindeer to it.

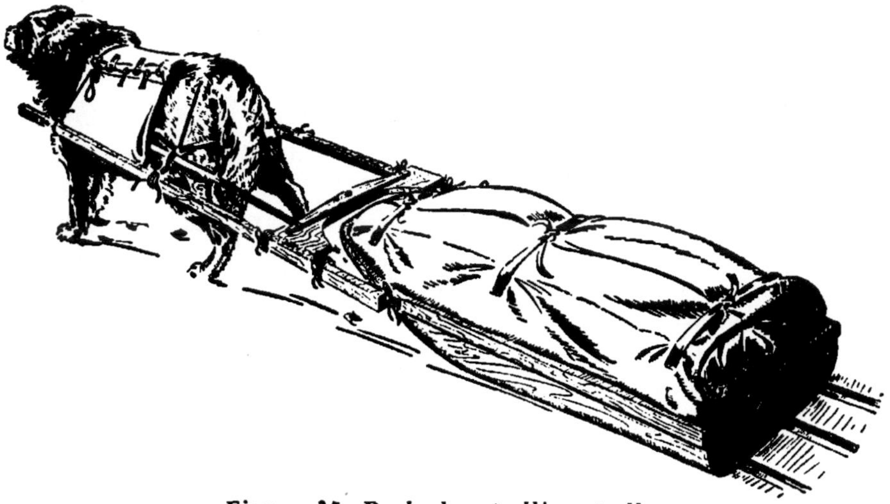

Figure 35. Pack dog pulling pulka.

(2) *Advantages.* The pulka has many advantages when compared with a flat toboggan:

(*a*) When used in deep, soft snow, the streamlined, boatlike sides produce less friction than a tarp-wrapped load on a flat toboggan.

(*b*) The extra steel shoeing on the three runners prevents bottom wear, holds the wooden surface of the pulka off hard, ice-crusted surfaces and helps greatly to prevent slewing.

(*c*) The U-shaped keel also helps to prevent slewing.

(*d*) The boatlike construction makes it possible to keep contents absolutely dry when crossing small streams and patches of water. The pulka will float a 60-pound load with 1½ inches clearance above water.

(3) *Use.* A dog weighing 125 to 150 pounds can pull a load weighing up to 300 pounds in a pulka, depending on the térrain. When operating in mountainous country, the pulka should never be loaded with more than the dog's weight. On fairly flat country or on sea-ice with a fast surface, maximum loads may be hauled.

(4) *Care.* Ski-wax, or a thin coating of ice on the bottom of the pulka, reduces friction and makes the load much easier to pull.

Section II. PACK DOG TRAINING

92. BASIC TRAINING. The pack dog must learn to understand thoroughly and obey without hesitation the four basic commands.

a. "Back". (1) The pack dog is not taught to "heel" (to remain at his master's left knee), for in this position he would constantly bump the trainer with his pack in the summer, and step on his snowshoes in the winter. The correct position of the pack dog is behind his master, whether on the trail or in the open. He is therefore taught to respond to the command "back" by keeping behind his master.

(2) The most effective method of teaching this command is to take the dog out alone, at first on a leash and

later free, and to force him to remain in position behind the master by light taps on his nose with a small switch. Each time this correction is used, it should be preceded by the command "back" until the dog becomes familiar with the desired response to the command.

b. "Down". (1) The dog must learn to respond to this command instantly, without hesitation. It is preferable that he learn to respond to a silent hand signal (a raised hand); but it will be necessary to give him a verbal command in the early stages of training. The dog must be tested for obedience to this command in various situations—in the presence of game or other distractions, when he is at a considerable distance from the master, and at the onset of a fight. Many accidents can be avoided on the trail if all dogs are trained to respond immediately to the command "down".

(2) This command may be taught by repeating "down" while stepping on the dog's leash and gently forcing him to the ground. If a silent signal is given, the hand should be raised at the moment the command is given and while the dog is forced down.

c. "Stay". Once a dog is down, it is practical to have a way to make him remain in that position until ordered to change. This will allow the master to go about his business with the assurance that the dog will be in place when he looks for him. This is particularly important when unloading supplies or setting up camp.

d. "Come". The master may give the command "down", then the command "stay", go forward to examine some questionable area or route, and then call his dogs without having to return to them. The command "come", coupled with the name of the dog, may be given to bring up an individual dog which is lagging. Much of the value of a pack dog is determined by his obedience to this command. It is also good practice to train the dog to come when called by name or by a hand signal.

e. Refinements of these four commands and any additional commands that are necessary can be taught to the dogs if the master is an experienced dog handler. All such additional training, however, should be based upon the four basic commands and the actual work expected of the dog.

f. Various small devices may be used in the training and handling of pack dogs. Some trainers prefer a short piece of rubber hose or a small switch which will sting mildly, and others get along by using only the back of the hand. Individual dogs vary in their response to discipline; some will respond to a harsh verbal command alone and others require a tap with a switch. Regardless of what device is used, it must be remembered that pack dogs should *never* be severely whipped or beaten. Whips are absolutely forbidden in pack dog training. No chains should ever be used for punishment.

93. ADVANCED TRAINING. a. Group training. The average pack string is composed of four dogs, although an experienced trainer can handle as many as eight in one group. After each dog has received his basic training, and once his capabilities are fully familiar to the trainer, group training should start.

(1) Four dogs should be selected who match in weight, build, vitality, and speed. They should be chosen with the expectation that they will work and stay together.

(2) When the dogs have been selected, they should be worked together in pairs, putting different pairs together each day. If no difficulties arise, their training as a group may then be started. Training in the four basic commands must be continuous, whether the dogs work singly, in pairs, or in a group.

(3) Once the group is outside the kennel area, they should be let off the leash. The group must be trained to work freely at all times, for they must have a flexibility which is possible only if they are off leash. It is impractical to work the dogs fastened to a single line; if one dog has to stop for a moment, the whole group will have to stop, and this lack of freedom will be both tiring and irritating to the whole group. If the dogs have to travel over rough trail or bushwack, freedom and flexibility will be essential.

b. Training to the pack. Few dogs take naturally to

the pack. The first reaction often is to roll over, to try to shake the pack off, to chew it, or to lie down and cower.

(1) From the first day of basic training throughout the entire period of the dog's work, the pack should always be put on before any training or trail work is done. The dog loves to get away from his kennel and get the exercise and "freedom" of training work. If he always wears a pack under these circumstances, it will become a symbol of liberty and the dog will look forward to work with eagerness. It is very easy for a trainer to neglect to put on the pack during basic training and such negligence is a serious training error. The only time it is permissible to leave the pack off is when a sore or injury may be aggravated by its use.

(2) The pack should be perfectly fitted to the individual dog. After it has been fitted, the dog should be led around wearing the empty pack to see what his reaction is. For at least the first week, the pack should be loaded with two burlap sacks filled with straw. The weight will be negligible, but the bulk will begin the dog's training in handling himself and his pack when going through brush and timber.

(3) After the first week, the trainer may begin to add weight to the pouches. Ten pounds will usually be enough to start with. Later, increases in weight should be made in accordance with the ability of the dog to carry the load without excess strain. There should be a gradual conditioning of the dog to loads up to the maximum of 40 to 60 pounds. Some dogs will be able to pack their maximum loads after the third or fourth week, and others will require much longer before they can be considered fully conditioned.

(4) There should be no petting or playing after the pack has been put on the dog. The dog must realize that when he is wearing the pack he is expected to work. This does not mean that there may not often be a good deal of understanding and affection between the master and his dogs; however, when the dogs are working, it should be a businesslike understanding. Too much

petting while under pack will make it difficult for the master to enforce the basic commands.

(5) Activities of the trainers should be scheduled to permit the dogs to be taken out every day. Training should not be spasmodic; it will be much facilitated, particularly in the beginning, if the dogs realize they are to work regularly every day.

(6) When the dogs have completed their basic and advanced training, they will be able to pack a full load all day. They should then be run through a varied course of fire-conditioning, involving rifle and machine-gun fire, land mines, and, if possible, cannon fire which will produce concussion as well as noise. The dogs should first be exposed to fire at a distance, and later more closely, so that they will learn to respond to commands under fire without being distracted. The dogs should be conditioned to firing from all directions and to firing by others than the trainer.

INDEX

○

☆ 599354—1944

Made in United States
North Haven, CT
04 November 2026

21770789R00088